Nanoengineering Materials for Biomedical Uses

Emilio I. Alarcon · Manuel Ahumada
Editors

Nanoengineering Materials for Biomedical Uses

 Springer

Editors
Emilio I. Alarcon
Division of Cardiac Surgery
University of Ottawa Heart Institute
Ottawa, ON, Canada

Manuel Ahumada
Center for Applied Nanotechnology
Universidad Mayor
Santiago, RM, Chile

ISBN 978-3-030-31263-3   ISBN 978-3-030-31261-9   (eBook)
https://doi.org/10.1007/978-3-030-31261-9

This Springer imprint is published by the registered company Springer Nature Switzerland AG
The registered company address is: Gewerbestrasse 11, 6330 Cham, Switzerland

Preface

Nanomaterials and nanotechnology have helped to shape what we know as modern medicine. Unrebuttably, we are living in an era of brilliant medicine with advances in pretty much every field of medicine ranging from prevention all the way to artificial organs. However, as our population lives longer, the spectra and abilities of the therapeutics and diagnoses must be accordingly updated. Nanomaterials have opened a whole new world of possibilities for developing more advanced and precise diagnostic tools, drug delivery nanocarriers, and imaging and sensor elements for diagnostics. The superior physical and chemical properties of nanomaterials when compared to their bulk counterpart make the field of nanomaterials and nanotechnology a unique niche for developing novel therapeutics and diagnostic tools for translational medicine.

With nanomaterials having become so varied and their utilization in the field of biomedicine even more so, the current work documents and discusses each of these in a practical manner for fellow scientists and clinicians alike, as well as interdisciplinary scientists willing to enter the field. The simplicity of the layout is a key aspect contributing to its usefulness. This effort is to serve as an up-to-date library to inform readers about current research and the accompanying challenges, as well as inspire new enquiries.

Authored by nanomaterials researchers, this book covers recent advances in biomedical applications of nanomaterials. It begins with important fundamental information on nanomaterial synthesis and characterization and then dives into the latest advances of nanomaterial applications in the clinic, specifically in organ and tissue repair and regeneration. Overall, readers will gain an appreciation for current and relevant methods for synthesizing and characterizing nanomaterials along with an understanding of computational methods used for nanomaterial design, as well as an in-depth look into the applications of nanomaterials for organ repair. Last, the regulation of nanomaterials is discussed, a topic that is still developing with the field, trying to incorporate the diversity of the field while not limiting the innovation.

Dr. Emilio I. Alarcon, Ph.D., MRSC
Book Editor
Principal Investigator and Laboratory Director
Division of Cardiac Surgery
University of Ottawa Heart Institute
Ottawa, ON K1Y 4W7, Canada

Assistant Professor, Department of Biochemistry,
Microbiology, and Immunology, Faculty of Medicine
University of Ottawa
Ottawa, ON K1H 8M5, Canada
http://www.beatsresearch.com

Dr. Manuel Ahumada, Ph.D.
Book Editor
Assistant Professor and Principal Investigator
Center for Applied Nanotechnology (CNAP), Faculty of Sciences
Universidad Mayor, Campus Huechuraba
Santiago, RM, Chile
http://cnap.umayor.cl

Erik Jacques, B.Sc.
Book Assistant Editor
University of Ottawa Heart Institute
Ottawa, ON K1Y 4W7, Canada
http://www.beatsresearch.com

Caitlin Lazurko, M.Sc.
Book Assistant Editor
Department of Biochemistry, Microbiology, and Immunology
Faculty of Medicine, University of Ottawa
Ottawa, ON K1H 8M5, Canada
http://www.beatsresearch.com

Acknowledgements

Editor: Emilio I. Alarcon, Ph.D., thanks to the financial support of the Canadian Institutes of Health Research (CIHR), the Natural Sciences and Engineering Research Council of Canada (NSERC), the Ministry of Economic Development, Job Creation, and Trade for an Early Researcher Award, and the New Frontiers in Research Fund—Exploration for a Research Grant.

Editor: Manuel Ahumada, Ph.D., thanks to the CONICYT–FONDECYT (Iniciación en la Investigación) grant #11180616 and to FDP, Universidad Mayor, grant #I-2019077, for financial support.

Assistant Editor: Caitlin Lazurko, M.Sc., thanks to the Queen Elizabeth II Graduate Scholarships in Science and Technology for financial support.

Contents

Contributors

Gabriel Abarca Facultad de Ciencias, Centro de Nanotecnología Aplicada, Universidad Mayor, Santiago, Chile

Manuel Ahumada Facultad de Ciencias, Centro de Nanotecnología Aplicada, Universidad Mayor, Huechuraba, RM, Chile

Emilio I. Alarcon Division of Cardiac Surgery, University of Ottawa Heart Institute, Ottawa, ON, Canada;
Department of Biochemistry, Microbiology, and Immunology, Faculty of Medicine, University of Ottawa, Ottawa, Canada

Angela Auriat Ottawa Hospital Research Institute, Ottawa, ON, Canada

Isabelle Brunette Department of Ophthalmology, Faculty of Medicine, Université de Montréal, Montréal, QC, Canada;
Centre de Recherche, Hôpital Maisonneuve-Rosemont, Montréal, QC, Canada

Xudong Cao Faculty of Engineering, University of Ottawa, Ottawa, ON, Canada

Suzan Chen Ottawa Hospital Research Institute, Ottawa, ON, Canada

Maria DeRosa Department of Chemistry and Institute of Biochemistry, Carleton University, Ottawa, ON, Canada

Jesum Alves Fernandes School of Chemistry, University of Nottingham, Nottingham, UK

Walfre Franco Wellman Center for Photomedicine, Massachusetts General Hospital, Boston, MA, USA;
Department of Dermatology, Harvard Medical School, Boston, MA, USA

May Griffith Department of Ophthalmology, Faculty of Medicine, Institute of Biomedical Engineering, Université de Montréal, Montréal, QC, Canada; Centre de Recherche, Hôpital Maisonneuve-Rosemont, Montréal, QC, Canada

François-Xavier Gueriot Department of Ophthalmology, Grenoble Alpes University, La Tronche, Grenoble, France

Cristián Gutiérrez-Cerón Facultad de Química y Biología, Universidad de Santiago de Chile, Santiago, Chile

Erik Jacques Division of Cardiac Surgery, University of Ottawa Heart Institute, Ottawa, ON, Canada

Anna Koudrina Department of Chemistry and Institute of Biochemistry, Carleton University, Ottawa, ON, Canada

Caitlin Lazurko Division of Cardiac Surgery, University of Ottawa Heart Institute, Ottawa, ON, Canada; Department of Biochemistry, Microbiology, and Immunology, Faculty of Medicine, University of Ottawa, Ottawa, Canada

Janani Mahendran Department of Chemical and Biological Engineering, University of Ottawa, Ottawa, ON, Canada

M. Andrea Molina Torres Departamento de Química Orgánica, Facultad de Ciencias Químicas, Universidad Nacional de Córdoba, Córdoba, Argentina; Consejo Nacional de Investigaciones Científicas y Técnicas (CONICET), INFIQC, Córdoba, Argentina

Rodrigo N. Núñez Departamento de Química Orgánica, Facultad de Ciencias Químicas, Universidad Nacional de Córdoba, Córdoba, Argentina; Consejo Nacional de Investigaciones Científicas y Técnicas (CONICET), INFIQC, Córdoba, Argentina

Natalia L. Pacioni Departamento de Química Orgánica, Facultad de Ciencias Químicas, Universidad Nacional de Córdoba, Córdoba, Argentina; Consejo Nacional de Investigaciones Científicas y Técnicas (CONICET), INFIQC, Córdoba, Argentina

Horacio Poblete Center for Bioinformatics and Molecular Simulation, Universidad de Talca, Talca, Chile; Millennium Nucleus of Ion Channels-Associated Diseases (MiNICAD), Universidad de Talca, Talca, Chile

Marc Ruel BEaTS Research Program, Division of Cardiac Surgery, University of Ottawa Heart Institute, Ottawa, Canada

Veronika Sedlakova BEaTS Research Program, Division of Cardiac Surgery, University of Ottawa Heart Institute, Ottawa, Canada

Fiona Simpson Department of Ophthalmology, Faculty of Medicine, Institute of Biomedical Engineering, Université de Montréal, Montréal, QC, Canada; Centre de Recherche, Hôpital Maisonneuve-Rosemont, Montréal, QC, Canada

Jean-Philippe St-Pierre Department of Chemical and Biological Engineering, University of Ottawa, Ottawa, ON, Canada

Erik J. Suuronen BEaTS Research Program, Division of Cardiac Surgery, University of Ottawa Heart Institute, Ottawa, Canada

Eve C. Tsai Ottawa Hospital Research Institute, Ottawa, ON, Canada; Division of Neurosurgery, Suruchi Bhargava Chair in Spinal Cord and Brain Regeneration Research, University of Ottawa, Ottawa, ON, Canada

Ariela Vergara-Jaque Center for Bioinformatics and Molecular Simulation, Universidad de Talca, Talca, Chile; Millennium Nucleus of Ion Channels-Associated Diseases (MiNICAD), Universidad de Talca, Talca, Chile

Ying Wang Wellman Center for Photomedicine, Massachusetts General Hospital, Boston, MA, USA; Department of Dermatology, Harvard Medical School, Boston, MA, USA

Ricardo A. Zamora Facultad de Química y Biología, Universidad de Santiago de Chile, Santiago, Chile

Matías Zúñiga Center for Bioinformatics and Molecular Simulation, Universidad de Talca, Talca, Chile

Chapter 1
Nanomaterials for Its Use in Biomedicine: An Overview

Caitlin Lazurko, Erik Jacques, Manuel Ahumada and Emilio I. Alarcon

Abstract The rapid incorporation of nanostructures in regenerative medicine can be considered one of the biggest leaps in the production of novel materials for repair and regeneration of damaged tissues. However, despite a large number of articles published, clinical use of these materials is still in its infancy. The complexity and interdisciplinary nature of research aimed to repair damaged tissue and failing organs are the main limiting factors that have halted the progression for developing novel structures for tissue repair. In the present chapter, we revise fundamental concepts to be considered when designing technologies that will have to undergo scrutiny by regulatory agencies prior to being used in humans.

1.1 Introduction

Modern medicine relies on functional materials to provide tools which allow the partial, or even more desirable, the complete restoration of the functionality of damaged organs and tissues. Paradoxically, the increase in life expectancy and improved surgical outcomes presents a new challenge for developing novel materials for organ repair. Thus, what was considered a significant achievement in tissue engineering in the past, such as the first human donor cornea transplant, has become a routine procedure. However, cornea transplantation is limited by donor shortage and graft rejection in chronically inflamed eyes (see Chap. 8). Thus, novel therapeutics in the field of corneal tissue repair needs to circumvent the worldwide shortage while

C. Lazurko · E. Jacques · E. I. Alarcon (✉)
Division of Cardiac Surgery, University of Ottawa Heart Institute,
40 Ruskin Street, Ottawa, ON K1Y 4W7, Canada
e-mail: Ealarcon@ottawaheart.ca

C. Lazurko · E. I. Alarcon
Department of Biochemistry, Microbiology, and Immunology, Faculty of Medicine,
University of Ottawa, Ottawa, ON K1H 8M5, Canada

M. Ahumada
Facultad de Ciencias, Centro de Nanotecnología Aplicada, Universidad Mayor,
Huechuraba, RM, Chile

© Springer Nature Switzerland AG 2019
E. I. Alarcon and M. Ahumada (eds.), *Nanoengineering Materials for Biomedical Uses*,
https://doi.org/10.1007/978-3-030-31261-9_1

providing implants capable of modulating chronic inflammation. The level of complexity for engineering tissues become more challenging in highly perfused and contractile organs, as is the case of the heart muscle, where materials must also incorporate electroconductive moieties (see Chap. 9). Synchronic conductivity and alignment are also of prime importance in regenerating nerves (Chap. 7). Considering soft tissues, for example, the skin which is the largest organ in the human body and the primary target of external insults; nowadays developing functional biomaterials for skin repair requires pushing the boundaries of materials chemistry for producing novel biologically compatible templates that allow functional skin regeneration with minimal scarring (see Chap. 6). This push in materials with improved biological properties becomes even more challenging for tissues that will be exposed to high shear forces such as articular cartilage (Chap. 5). Alongside the exponential growth in knowledge surrounding the underlying mechanisms involved in wound healing and tissue regeneration during the last two decades, there has been an evident need for novel strategies and therapeutics for tissue repair. This new body of literature, however, is not enough for us to fine tune the biophysical properties of the biomaterials to make them better "at healing". Thus, incorporating nanoscale components as structural building blocks for modulating the biophysical properties of the materials, which will ultimately allow the manipulation of cell-matrix interactions (Fig. 1.1).

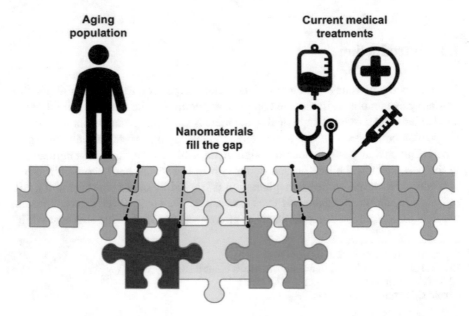

Fig. 1.1 The role of nanomaterials is to fill the unmet needs in the field of medicine. These materials can be used to modify the biophysical properties of biomaterials, control cell-matrix interactions, and revolutionize the field of tissue engineering and regenerative medicine, especially with the aging population and increased medical demands

In the following sections of this chapter, we will briefly revise the history of biomaterials alongside with fundamental principles of nanotechnology and regenerative medicine.

1.2   Brief History of Materials Used in Medicine

The term nanotechnology was first introduced in 1974 by Taniguchi to describe the engineering of nanoscale materials [1, 2]. However, nanomaterials have been present in human history since ancient times, when colloidal solutions of gold nanoparticles were used to dye glass [3, 4]. The Lycurgus Cup, an example from the fourth century A.D., used nanoparticulated metal dispersed in glass to give color to the cup, and the color changed depending on the light incidence angle [3–5]. Nanotechnology advances in the last decades have provided scientists with tools to investigate, engineer, and control assemblies of atoms and molecules less than 100 nm in size [6, 7]. As nanotechnology has progressed, its nature has exponentially diversified, becoming an intrinsically interdisciplinary field, where understanding the nanoscale interactions are essential for developing new technologies and therapies [1, 4]. In the 1990s, the term nanomedicine started to be used to refer to nanomaterials with potential medical applications [4, 6–9]. Today, nanomedicine is often subdivided into either the development of tools for medical diagnosis and therapies or fundamental research on understanding interactions and interface between chemical, biological, and physical sciences [1, 9].

Early applications of nanomaterials in medicine were often completed without a deep understanding of the interactions at the nanoscale level. Nanomaterials were used without the devices and technologies available today, such as electron microscopy, to be able to identify the importance of the nanoscale size of the materials and the nanoscale interactions that were occurring. For example, in the nineteenth century, nanoporous ceramic filters were used to separate viruses [4]. Advancements in microscopy led to a better understanding of cell structures and interactions, and further microscopy development including the development of atomic force microscopy and the scanning tunnel microscope resulted in the ability to visualize objects at the nanometer scale [4]. It was these advancements in technology that allowed the field of nanotechnology and nanomedicine to boom [4]. During the 1990s, tissue engineering had a boost when it merged with stem cell transplantation to become a much more influential field also known as regenerative medicine (William Haseltine would later coin the term in 1999) [10–12]. As products began to be successfully commercialized, the interest of the private sector also increased, which catalyzed the development and testing of a large variety of biomaterials [13]. However, the excitement was rather short-lived, as scientists tried to copy tissue formation rather than seek to understand the underlying mechanisms for tissue repair [11, 13, 14]. As a result, products that showed great promise in the lab failed, and coupled with the Y2K crash, meant that by 2002 the value of the industry was down by 90% [13, 15]. Out of the 20 FDA-approved products during that time, none remain on the market today [15].

Presently, the field has recovered from that crash and is now much more diversified [16–18]. There has been a switch in focus to simpler acellular products such as biomaterials, and a continued search for other avenues of inquiry, such as nanotechnology, which are actively being introduced into the field of medicine for a range of applications including drug delivery, tissue engineering, diagnostics, therapies, and imaging [1, 7, 9, 19–21]. There has also been an increase in nanomaterial funding worldwide, with over $7 billion per year is being allocated to nanotechnology. The United States is leading the way in nanotechnology funding, which has increased since the signing of the twenty-first century Nanotechnology Research and Development Act (NRDA) in 2003. Many other countries, including the EU, Japan, and South Korea, are following suit and prioritizing research and development of nanotechnologies for various applications [1, 22].

With the increased interest in nanomaterial research, one big question that remains is the potential impact on human health. There are concerns that the unique properties of nanomaterials, which are discussed below, may have a negative impact on human health and the environment [1]. There is a lack of information regarding how nanomaterials interact with the world and their impact on the food chain. Moreover, as nanoparticles vary significantly in size, shape, and composition, their toxicity varies as well, with certain particles being known to be biocompatible and non-toxic, while others showing cell toxicity [23–25]. A collaborative approach should be taken by researchers when designing and testing nanoparticles to ensure they are designed to be effective for their application while remaining biocompatible [1]. There are also ethical questions concerning nanomaterials, including who benefits from and who controls the use of these technologies. Due to the novelty and diversity of nanomaterials and their applications in medicine, it is important to get a complete understanding of the benefits and risks associated with these materials before testing in vivo to ensure the safety of these technologies.

1.3   Fundamental Concepts on Nanomaterials

Some of the fundamental concepts surrounding nanomaterials that must be considered when designing nanomaterials, especially those for medical applications, are discussed here. Nanostructures are typically prepared by either a "top-down" or "bottom-up" method. The top-down method starts with the bulk material and follows a synthetic route to obtain the nanostructure. On the other hand, the bottom-up method starts with atoms and makes them coalescent to form nanostructures. Nonetheless, independent of the chosen route, the final product will have the same nanostructure properties, which will have different physical-chemical properties from those found in the original bulk material. Moreover, the material(s) that form a nanostructure can come from a variety of sources, being either biological or chemical in nature. For instance, metal nanostructures are famous among biomedical applications due to their tunable physical-chemistry, antibacterial, and biocompatibility properties. Popular choices are gold, silver, titanium, and copper [26–30]. Synthetic polymers are also a source of material for nanostructures, where they can be used to form nanoparticles, mesh-like composites, foams, among others [31]. Common synthetic polymers are poly-ethylene glycol and its derivates, poly-caprolactone and poly-vinyl alcohol, to name a few [32]. Similarly, natural polymers such as polypeptide chains, proteins, and carbohydrates and their derivates have also being used for nanostructure development [33]. Usually mentioned are collagen, fibrin, alginate, chitosan, and gelatin [34]. Nonetheless, proteins by themselves are nanostructures with potential biomedical applications [35]. While they are ubiquitous, their potential as drug nanocarriers has been widely explored, with remarkable cases, such as the use of serum albumin (either from human or bovine sources) [36].

The keystone for the explosion in nanomaterial applications, particularly for those of synthetic nature, lies in the fact that these nanostructures have properties that vary from those of the bulk material [1, 6]. Most of the properties of macroscopic materials are described, unequivocally, by classical physics, which is based on empirical science at the macro-scale. However, nanomaterials, as aforementioned, respond to a different size scale, which dramatically changes the way the physics works. In the early 1900s, the term quantum physics started being introduced from the theoretical field, where later experimental physicists probed the existence of this new branch, that differs entirely from its classical counterpart. Particularly in the case of nanomaterials for biomedical applications, the high surface area along with quantum effects results in unique optical, magnetic, and electronic properties [1, 5, 6, 20, 21].

First, the available surface area is one of the most relevant properties of nanomaterials, independent of their origin or shape [37]. This allows nanomaterials to adsorb different particles, especially proteins and drugs, onto their surface. These molecules bound to the surface can then impact nanoparticle stability, solubility, biocompatibility, and its interactions with other molecules in their environment [20, 21]. Their surface and composition can also be modified to match the environment of the tissue they are interacting with, in a process called surface nano-engineering.

The second relevant feature of nanostructures corresponds to the shape, where virtually any shape can be considered a nanostructure, as long as the structure fulfills the conditions, vide supra. Thus, a wide range of shapes can be found in the literature, including spheres, rods, cubes, tubes, flower, cage, foam, flake, ring, mesh, amorphic [38]. Despite the number of shapes mentioned, the access to those is limited by several factors such as synthesis method, components, and experimental conditions. Furthermore, the target application for nanostructure use also plays a fundamental role in the shape selection; for example, when considering nanoparticles for their use in the near-infrared section of the spectrum, usually, spherical nanoparticles would not present a plasmon response (no absorption), however, the elongation into a rod-like shape for the same nanostructure will increase the longitudinal plasmon promoting the generation of a signal in the near-infrared region. Therefore, it is important to consider the application of the nanomaterial when choosing the structure.

Next, the specific optical, magnetic, and electrical properties of the nanostructure stand out, particularly for metal nanostructures, from other materials when designing new biomedical technologies. Two main characteristics describe these phenomena;

first, electrons are distributed differently in the system and second, the nanostructures interact with light in a unique manner. When in a bulk material, electrons can be described as a continuum; however, in the case of nanostructures, electrons have a discontinued behavior, which can likely be controlled. Second, since the nanostructures have smaller sizes, they interact with light, especially wavelengths generally used for biomedicine (UV-A to NIR), in a different manner than the bulk material. The properties mentioned here will be further explored and expanded in the following chapters of this book.

The properties of nanostructures, as mentioned above, can improve the biocompatibility of the materials and alter the interactions of the materials with the host environment [19, 20]. Nanoparticles can also modify the micro-environment in which they are present, which can influence the cell's fate. They can be used to enhance interactions with the host and can be used to engineer biomaterials that more closely mimic native tissue and endogenous conditions [19]. The versatility of nanomaterials extends beyond the materials they are synthesized from and the particles used to coat their surface, as nanomaterials can be applied in a variety of different manners. For example, they can be used as a thin coating on surfaces, such as in electronics or on prosthetic implants, they can be embedded in a material, such as a biomatrix, or free nanoparticles can be used [1].

1.4   Brief Considerations for Regenerative Medicine

Today, tissue loss of function can generally only be solved with organ and tissue transplantation [10, 11]. However, donor availability is scarce, and the demands of the aging population and its chronic diseases are ever-growing [10, 11]. In the 1960s, the limitations of transplantation began to be felt as chronic diseases were on the rise [14, 17, 39, 40]. Concurrently, scientists such as Alex Carrel began culturing cells and thus were beginning to grow and keep tissues alive in vitro [12]. The processes of degeneration and regeneration were now being studied. It was not until the 1980s, in Boston, Massachusetts, where Dr. Joseph Vacanti and Robert Langer decided to use this knowledge to create in vitro grown skin grafts (Epicel® and Apligraf®) [12, 14]. Now everyone was working on trying to create skin or cartilage grafts. This is where regenerative medicine (RM) comes into play (also referred to as tissue engineering and regenerative medicine, or TERM for short).

Surprisingly, the concept of tissue regeneration began in myths, where a common example is the Greek myth of Prometheus, a Titan who received a terrible fate from Zeus after having gifted humanity with fire "*He was bound to a rock where an eagle would feast on his liver every day, and every night said liver would regenerate leading to an endless loop of torture* [41]." The idea of regeneration persisted through the millennia until the twentieth century, where RM came into fruition. Greenwood et al. stated:

Regenerative medicine is an emerging interdisciplinary field of research and clinical applications focused on the repair, replacement or regeneration of cells, tissues or organs to restore impaired function resulting from any cause, including congenital defects, diseases, trauma and aging. It uses a combination of several technological approaches that moves it beyond traditional transplantation and replacement therapies. These approaches may include, but are not limited to, the use of soluble molecules, gene therapy, stem cell transplantation, tissue engineering and the reprogramming of cell and tissue types. [42]

RM is a branch of biomedical science that uses various strategies to restore function to damaged or diseased human tissue and tries to regenerate lost tissue and/or organs. This has led to immense scientific, private (allied market research estimates that the market for RM will be worth $67.5 billion dollars by 2020), and media interest, which refers to it as "the most promising healthcare technology ever put forward" [12, 40]. The regenerative therapies are leading a paradigm shift from treatment-based to cure-based therapies which will have a profound impact not only on the quality of healthcare but also on its economics as the financial burden of chronic diseases would be significantly lifted [17, 40, 43].

As mentioned in Greenwood's definition, RM's arsenal is vast, and since 2006, it has increased to include bioreactors, bioprinting, and nanotechnology [42]. While RM's focus since the early 2000s has been on the use of human stem cells, this focus has shifted significantly to the use of acellular products either concurrently or without cells for tissue regeneration [10, 12, 17]. In particular, this refers to controlled release matrices and scaffolds; materials where the principles of nanotechnology are being regularly used. These regenerative therapies can currently be divided into three categories: allogeneic, autologous, and scaffolds. Allogenic therapies are cell therapies that use a universal donor cell; autologous therapies utilize donor cells harvested from the patient; and scaffolds include the use of decellularized extracellular matrices (ECMs) or synthesized biomaterials [39, 44, 45]. The authors do recognize that hybrid models exist and that bioprinting could also be considered a category but since they are young strategies, they have yet to be included as such.

RM seems theoretically promising, however, the results in clinic have not reflected this. Currently, all cell therapies remain experimental, except for hematopoietic stem cell transplants, simultaneously, acellular products have had little success making it to market; it is also a mainstream opinion that cell therapies have shown little efficacy and that, to date, RM has underperformed [12, 45, 46]. The reasons stem from the field's novelty, where it has created many challenges. Pre-clinically, scientists have yet to fully understand the regenerative mechanisms behind their therapies [10]. Clinically, the tumorigenicity, immunogenicity, and risks of the procedure delivery are all unsolved obstacles involved with cell therapies [46]. Additionally, regenerative therapies are meant to be implanted and remain with the patient for a prolonged period. Unfortunately, long-term follow-up studies do not exist for clinical trials which makes it difficult to ensure regulatory agencies and the public of the safety and efficacy of the therapy [40, 43]. Post-clinic, there is difficulty in identifying the proper business model for companies hoping to enter the RM market [12]. The regulatory and reimbursement policies for such novel technologies have been difficult for countries to determine [10]. However, the most significant challenges are the ones related to manufacturing. The automation and scale-out strategies of the manufacturing process to reduce cost, contamination, and human error do not exist for such complex biological therapies [40, 47]. As well, the industrialization technology simply does not exist [39]. Despite the turnaround, RM and its growing number of clinical trials are reaching a critical mass and becoming a major player into the biomedical field [12, 47]. Whether this will prosper, remains to be seen. Nonetheless, nanotechnology has played a role in the field's progression.

Thus, there is now also a high degree of optimism for regenerative therapies and once again a rush to get them through to clinical trials [45]. Governments now recognize RM as being at the forefront of healthcare and institutions dedicated to its practice have increased over the past decade [39]. As mentioned above, while the discoveries made on the bench-side are ever-increasing (such as the discovery of induced pluripotent stem cells for example), success on the bench-side is still lacking [18, 48]. Therefore, clinical trials have been increasing but are proceeding with caution [45]. Additionally, in accordance with the maturation of the field, various attempts to rectify the challenges already discussed have been made, for example: companies such as Canada's Centre for Commercialization of Regenerative Medicine have been created to help researchers (academic or private) facilitate the translation of their therapies by decreasing risk during the development phase; the Mayo Clinic has created a theoretical blueprint for the "discovery, translation and application of regenerative medicine therapies for accelerated adoption into standard of care"; and legislation in places such as the United States, the EU, and Japan has been passed to allow accelerated conditional approval of RM technologies so as to be more readily available to the public [39, 47, 49–51].

1.5   Outlook and Future Perspectives

Nanotechnology allows for the production of efficient markers and extremely precise diagnostic tools and imaging devices, which allows for early diagnoses, all of which can improve treatments and quality of life for patients and decrease overall morbidity and mortality rates. These devices are also in line with regenerative medicine in that they help improve our understanding of interactions in the human body which allows for the development of new therapies [52]. Understanding the pathophysiological basis of diseases and how nanomaterials interact with cells and tissues in the body are essential to the design, development, and application of nanomaterials in medicine [20, 21]. There is currently a gap in knowledge surrounding nanomaterial interactions in the human body, including, toxicity, pharmacokinetics, and pharmacodynamics, which limits the technologies used and developed today [6]. However, nanomaterials have the potential to improve personalized medicine as well as the targeting of therapeutics, dose-response, and bioavailability, among many other aspects of medicine [6]. They show promise in the development of the multifunctional and next generation of biomedical devices that will further improve healthcare [6, 19]. Moreover, the broad range of nanomedicine to include genetics, molecular biology, cellular biology, chemistry, biochemistry, material science, proteomics, and bioengineering means that advances in this field will have broad applications in field of science and greatly improve patient care [6]. Overall, nanomaterials hold great promise for medical applications, and many avenues for nanomaterial application have yet to be explored.

Acknowledgements Dr. Alarcon thanks the Canadian Institutes of Health Research (CIHR), the Natural Sciences and Engineering Research Council of Canada (NSERC), the support of the Ministry of Economic Development, Job Creation and Trade for an Early Researcher Award, and the New Frontiers in Research Fund—Exploration for a research Grant. Ms. Lazurko thanks the Queen Elizabeth II Graduate Scholarships in Science and Technology for financial support. Dr. Ahumada thanks the CONICYT—FONDECYT (Iniciación en la Investigación) grant #11180616.

Disclosure All authors have read and approved the final version.

References

1. Nanoscience and nanotechnologies: opportunities and uncertainties. London, UK: The Royal Society and The Royal Academy of Engineering; 2004.
2. Taniguchi N. On the basic concept of nanotechnology. In: Proceedings of the international conference on production engineering. Tokyo, Japan: Japan Society of Precision Engineering; 1974.
3. Goesmann H, Feldmann C. Nanoparticulate functional materials. Angew Chem. 2010;49(8):1362–95.
4. Krukemeyer MGKV, Huebner F, Wagner W, Resch R. History and possible uses of nanomedicine based on nanoparticles and nanotechnological progress. J Nanomed Nanotechnol. 2015;6(6):336.
5. Stamplecoskie K. Silver nanoparticles: from bulk material to colloidal nanoparticles. In: Alarcon EI, Griffith M, Udekwu KI, editors. Silver nanoparticle applications: in the fabrication and design of medical and biosensing devices. Cham: Springer International Publishing; 2015. p. 1–12.
6. Ventola CL. The nanomedicine revolution: part 1: emerging concepts. P T. 2012;37(9):512–25.
7. Lehner R, Wang X, Marsch S, Hunziker P. Intelligent nanomaterials for medicine: carrier platforms and targeting strategies in the context of clinical application. Nanomed Nanotechnol Biol Med. 2013;9(6):742–57.
8. Bangham AD, Standish MM, Watkins JC. Diffusion of univalent ions across the lamellae of swollen phospholipids. J Mol Biol. 1965;13(1):238-IN27.
9. Wagner V, Dullaart A, Bock A-K, Zweck A. The emerging nanomedicine landscape. Nat Biotechnol. 2006;24:1211.
10. Sampogna G, Guraya SY, Forgione A. Regenerative medicine: historical roots and potential strategies in modern medicine. J Microsc Ultrastruct. 2015;3(3):101–7.
11. Slingerland AS, Smits AIPM, Bouten CVC. Then and now: hypes and hopes of regenerative medicine. Trends Biotechnol. 2013;31(3):121–3.
12. Kaul H, Ventikos Y. On the genealogy of tissue engineering and regenerative medicine. Tissue Eng Part B. 2015;21(2):203–17.
13. Berthiaume F, Maguire TJ, Yarmush ML. Tissue engineering and regenerative medicine: history, progress, and challenges. Annu Rev Chem Biomol Eng. 2011;2(1):403–30.
14. Kemp P. History of regenerative medicine: looking backwards to move forwards. Regen Med. 2006;1(5):653–69.

15. Lysaght MJ, Hazlehurst AL. Tissue engineering: the end of the beginning. Tissue Eng. 2014;383(9913):193–5.
16. Lysaght MJ, Jaklenec A, Deweerd E. Great expectations: private sector activity in tissue engineering, regenerative medicine, and stem cell therapeutics. Tissue Eng Part A. 2008;14(2):305–15.
17. Mason C, Dunnill P. A brief definition of regenerative medicine. Regen Med. 2008;3(1):1–5.
18. Chen C, Dubin R, Kim MC. Emerging trends and new developments in regenerative medicine: a scientometric update (2000–2014). Expert Opin Biol Ther. 2014;14(9):1295–317.
19. Bhat S, Kumar A. Biomaterials and bioengineering tomorrow's healthcare. Biomatter. 2013;3(3):e24717.
20. De Jong WH, Borm PJA. Drug delivery and nanoparticles:applications and hazards. Int J Nanomed. 2008;3(2):133–49.
21. Yao J, Yang M, Duan Y. Chemistry, biology, and medicine of fluorescent nanomaterials and related systems: new insights into biosensing, bioimaging, genomics, diagnostics, and therapy. Chem Rev. 2014;114(12):6130–78.
22. Bhushan B. Introduction to nanotechnology. In: Bhushan B, editor. Springer handbook of nanotechnology. Berlin, Heidelberg: Springer Berlin Heidelberg; 2017. p 1–19.
23. Dreher KL. Health and environmental impact of nanotechnology: toxicological assessment of manufactured nanoparticles. Toxicol Sci. 2004;77(1):3–5.
24. McLaughlin S, Ahumada M, Franco W, Mah TF, Seymour R, Suuronen EJ, Alarcon EI. Sprayable peptide-modified silver nanoparticles as a barrier against bacterial colonization. Nanoscale. 2016;8(46):19200–3.
25. Lam C-W, James JT, McCluskey R, Hunter RL. Pulmonary toxicity of single-wall carbon nanotubes in mice 7 and 90 days after intratracheal instillation. Toxicol Sci. 2004;77(1):126–34.
26. Ahumada M, Suuronen EJ, Alarcon EI. Biomolecule silver nanoparticle-based materials for biomedical applications. In: Martínez LMT, Kharissova OV, Kharisov BI, editors. Handbook of ecomaterials. Cham: Springer International Publishing; 2017. p. 1–17.
27. Burdusel AC, Gherasim O, Grumezescu AM, Mogoanta L, Ficai A, Andronescu E. Biomedical applications of silver nanoparticles: an up-to-date overview. Nanomaterials (Basel). 2018;8(9).
28. Fei Yin Z, Wu L, Gui Yang H, Hua Su Y. Recent progress in biomedical applications of titanium dioxide. Phys Chem Chem Phys. 2013;15(14):4844–58.
29. Goel S, Chen F, Cai W. Synthesis and biomedical applications of copper sulfide nanoparticles: from sensors to theranostics. Small. 2014;10(4):631–45.
30. Elahi N, Kamali M, Baghersad MH. Recent biomedical applications of gold nanoparticles: a review. Talanta. 2018;184:537–56.
31. Maitz MF. Applications of synthetic polymers in clinical medicine. Biosurf Biotribol. 2015;1(3):161–76.
32. Andonova V. Synthetic polymer-based nanoparticles: intellogent drug delivery systems. In: Reddy B, editor. Acrylic polymers in healthcare. London, UK: IntechOpen; 2017. p. 27.
33. Olatunji O. Biomedical application of natural polymers. In: Olatunji O, editor. Natural polymers: industry techniques and applications. Cham: Springer International Publishing; 2016. p. 93–114.
34. Aravamudhan A, Ramos DM, Nada AA, Kumbar SG. Chapter 4—Natural polymers: polysaccharides and their derivatives for biomedical applications. In: Kumbar SG, Laurencin CT, Deng M, editors. Natural and synthetic biomedical polymers. Oxford: Elsevier; 2014. p. 67–89.
35. Jain A, Singh SK, Arya SK, Kundu SC, Kapoor S. Protein nanoparticles: promising platforms for drug delivery applications. ACS Biomater Sci Eng. 2018;4(12):3939–61.
36. Kratz F. Albumin as a drug carrier: design of prodrugs, drug conjugates and nanoparticles. J Control Release. 2008;132(3):171–83.
37. Ahumada M, Lissi E, Montagut AM, Valenzuela-Henriquez F, Pacioni NL, Alarcon EI. Association models for binding of molecules to nanostructures. Analyst. 2017;142(12):2067–89.
38. Wei G, Ma PX. Nanostructured biomaterials for regeneration. Adv Funct Mater. 2008;18(22):3566–82.

39. Terzic A, Pfenning MA, Gores GJ, Harper CM Jr. Regenerative medicine build-out. Stem Cells Transl Med. 2015;4(12):1373–9.
40. Allickson JG. Emerging translation of regenerative therapies. Clin Pharmacol Ther. 2017;101(1):28–30.
41. Broughton KM, Sussman MA. Enhancement strategies for cardiac regenerative cell therapy. Circ Res. 2018;123(2):177–87.
42. Greenwood HL, Thorsteinsdottir H, Perry G, Renihan J, Singer P, Daar A. Regenerative medicine: new opportunities for developing countries. Int J Biotechnol. 2006;8(1 2):60–77.
43. Caplan AI, West MD. Progressive approval: a proposal for a new regulatory pathway for regenerative medicine. Stem Cells Transl Med. 2014;3(5):560–3.
44. Hunsberger J, Harrysson O, Shirwaiker R, Starly B, Wysk R, Cohen P, Allickson J, Yoo J, Atala A. Manufacturing road map for tissue engineering and regenerative medicine technologies. Stem Cells Transl Med. 2015;4(2):130–5.
45. Li MD, Atkins H, Bubela T. The global landscape of stem cell clinical trials. Regen Med. 2014;9(1):27–39.
46. Mount NM, Ward SJ, Kefalas P, Hyllner J. Cell-based therapy technology classifications and translational challenges. Philos Trans R Soc Lond B Biol Sci. 2015;370(1680).
47. Heathman TR, Nienow AW, McCall MJ, Coopman K, Kara B, Hewitt CJ. The translation of cell-based therapies: clinical landscape and manufacturing challenges. Regen Med. 2015;10(1):49–64.
48. Marincola FM. The trouble with translational medicine. J Intern Med. 2011;270(2):123–7.
49. Mason C, McCall MJ, Culme-Seymour EJ, Suthasan S, Edwards-Parton S, Bonfiglio GA, Reeve BC. The global cell therapy industry continues to rise during the second and third quarters of 2012. Cell Stem Cell. 2012;11(6):735–9.
50. Faulkner A. Law's performativities: shaping the emergence of regenerative medicine through European Union legislation. Soc Stud Sci. 2018;42(5):753–74.
51. Tobita M, Konomi K, Torashima Y, Kimura K, Taoka M, Kaminota M. Japan's challenges of translational regenerative medicine: act on the safety of regenerative medicine. Regen Med. 2016;4:78–81.
52. Singh M, Singh S, Prasad S, Gambhir IS. Nanotechnology in medicine and antibacterial effect of silver nanopartices. Digest J Nanomater Biostruct. 2008;3(3):115–22.

Chapter 2
Synthesis and Characterization of Nanomaterials for Biomedical Applications

Natalia L. Pacioni, M. Andrea Molina Torres and Rodrigo N. Núñez

Abstract This chapter aims to provide a critical overview of available synthetic methodologies for engineered nanomaterials for biomedical uses. We cover different kinds of nanoparticles with a focus on examples that have proven biocompatibility. Also, we included a summary of techniques and procedures for nanoparticle characterization. Finally, we discuss the remaining challenges in the preparation of nanomaterials for biomedicine.

2.1 Introduction

The last two decades have seen a growing interest in the medical application of nanomaterials (NMs) [1–3]. However, moving from synthesis in the laboratory to clinical use is the most challenging step, with most NMs being developed as proof of concept [2, 4]. By the end of 2017, the U.S. Food and Drug Administration (FDA) approved 50 new drugs containing nanoparticles, with only five of those using metal and metal oxide nanoparticles [5, 6].

It is important to bear in mind that NMs should have a series of characteristics to be used in biomedical applications [2], being:

- High monodispersity
- Water solubility
- Functionalization/Bioconjugation capability
- Stability under physiological conditions.

In this chapter, we will present the synthetic methodologies organized by the type of NM and covering only the cases that had led to biomedical uses. Further, we

N. L. Pacioni (✉) · M. A. Molina Torres · R. N. Núñez
Departamento de Química Orgánica, Facultad de Ciencias Químicas, Universidad Nacional de Córdoba, Haya de la Torre y Medina Allende s/n, X5000HUA, Ciudad Universitaria, Córdoba, Argentina
e-mail: nataliap@fcq.unc.edu.ar

Consejo Nacional de Investigaciones Científicas y Técnicas (CONICET), INFIQC, Córdoba, Argentina

© Springer Nature Switzerland AG 2019
E. I. Alarcon and M. Ahumada (eds.), *Nanoengineering Materials for Biomedical Uses*,
https://doi.org/10.1007/978-3-030-31261-9_2

exclude from this revision, particles derived from organic materials like proteins, peptides, liposomes, and polymers, whose synthetic protocols are well described in the literature [2].

There are different strategies to obtain engineered nanomaterials, using either *Top-down* or *Bottom-up* approaches [7, 8]. The latter method is more versatile, and most examples in this chapter belong to this classification.

Also, we include a summary of the most common characterization techniques to evaluate the physical and chemical properties of the NMs. We aim to provide the readers with key points and guidance to prepare biocompatible NMs.

2.2   Iron Oxide Nanoparticles

Currently, there are a few iron oxide nanoparticle (IONP) formulations approved by the FDA and the European Medicines Agency (EMA) for either the treatment of iron deficiency anemia or imaging applications, as well as some being clinically tested as therapies for thermal ablation of tumors [3, 5, 6]. Also, IONP and superparamagnetic IONP (SPIO) are of interest in the field of regenerative therapies for different tissues, including damaged cardiac [4].

Generally, the reported synthetic methodologies to prepare IONPs for biomedical applications involve two main steps, (i) synthesis of hydrophobic IONPs and (ii) modification of the IONP surface to make them hydrophilic. In this section, we present a summary of the protocols to prepare monodisperse and water-soluble IONPs. For an in-depth analysis of the advances on strategies to obtain biocompatible IONPs and ultra-small IONPs (<5 nm), the reader should further revise reviews [9, 10].

2.2.1   Synthesis of Spherical Fe₃O₄ Nanocrystals

Monodisperse magnetite nanoparticles (6–30 nm, Fig. 2.1) can be obtained using iron oxide powder [FeO(OH)] as a precursor, ground to 100–150 mesh and dissolved in oleic acid and 1-octadecene. In a three-neck flask equipped with a condenser and magnetic stirrer, the mixture is heated while stirring to 320 °C for a select time in an argon atmosphere. Pyrolysis of the iron carboxylate salt formed in situ leads to the formation of iron oxide nanocrystals. The combination of the iron precursor:oleic acid ratio and the reaction time tune the final nanoparticle size [11].

Another experimental approach uses iron (III) acetylacetonate [Fe(acac)₃] as the iron precursor to produce 6 nm magnetite nanocrystals [12]. Fe(acac)₃ is mixed with 1,2-hexadecanediol, oleic acid, and oleyl amine in a 1:5:3:3 ratio using benzyl ether (boiling point [b.p.] 298 °C) as the solvent, stirring, and under nitrogen flow. Once the mixture is heated to 200 °C for 2 h, under a blanket of nitrogen, it is refluxed at 300 °C for 1 h. Then, the reaction vessel is cooled, and ethanol is added to precipitate the IONPs. Purification involves several centrifugation and dissolution steps [12].

Fig. 2.1   TEM for Fe_3O_4 nanocrystals obtained with different average sizes of **a** 6.4 nm; **b** 19.9 nm; **c** 22.1 nm; and **d** 28.6 nm. The size was tuned using different FeO(OH):oleic acid ratios. Reprinted with permission from Ref. [11]. Copyright © 2004 Royal Chemical Society

Using phenyl ether (b.p. 259 °C) or 1-octadecene (b.p. 310 °C) leads to the formation of IONPs with average sizes of 4 and 12 nm, respectively. Variations in the amount of oleic acid, oleylamine, and solvent affect the resulting nanoparticle. The synthesis mechanism seems to involve the replacement of the "acac" ligand by the surfactants.

2.2.2   Synthesis of Iron Oxide Nanocubes

To obtain IONP (nanocubes), $Fe(acac)_3$ is dissolved in benzyl ether containing oleic acid and 4-phenylcarboxylic acid. The reaction is carried out under argon, heating to 290 °C at a rate of 20 °C/min. This temperature is maintained for 30 min, and then,

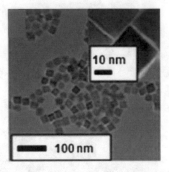

Fig. 2.2 TEM image of iron oxide nanocubes. Adapted with permission from Ref. [13]. Copyright © 2015 American Chemical Society

the solution is allowed to cool to room temperature. The resulting nanoparticles are precipitated using ethanol or acetone, followed by centrifugation and dispersion in chloroform. The side length of these nanocubes (Fig. 2.2) is approximately 22 nm [13].

2.2.3 Surface Modification

As mentioned above, the use of IONPs in biomedical applications requires water-soluble nanoparticles; however, most synthetic methodologies produce hydrophobic IONPs. Thus, after obtaining the IONP core, the modification or functionalization of the metal surface with hydrophilic compounds is needed. The main approaches to performing this modification are (1) surfactant addition and (2) surface surfactant exchange, as shown in Scheme 2.1. In (1), the result is a double-layer structure with the G moiety conferring water solubility; while in (2), the original surfactant is replaced by an amphiphilic surfactant that is capable of binding both the metal surface and the G group to provide the particles with a polar character [12].

For example, biocompatible IONPs were obtained by coating the nanoparticles with amphiphilic polymers [14] using a method reported previously for quantum dots [15] and based on the surfactant addition approach (Scheme 2.1). Briefly, a commercial triblock polymer composed of polybutylacrylate, polyethylacrylate, and polymethacrylic acid segments was derivatized with octylamine and stored in an ethanol-chloroform mixture. Then, the nanoparticles were encapsulated by the amphiphilic triblock polymer. After vacuum drying, the modified IONPs were dispersed in a polar solvent, and purification was performed by using a magnetic separator [14].

Also, water-soluble IONPs can be prepared by surfactant addition using the commercially available 1,2-Distearoyl-sn-glycero-3-phosphoethanolamine-N-[biotinyl(polyethyleneglycol)2000] (DSPE-PEG(2000)Biotin) [12]; 1,2-distearoyl-sn-glycero-3-phosphoethanolamine-N-[methoxy(polyethylene glycol)-2000] (DSPE-mPEG), or 1,2-distearoyl-sn-glycero-3-phosphoethanolamine-N-[amino(polyethylene glycol)-2000] (DSPE-PEG-NH$_2$) [13] shown in Fig. 2.3. In these cases, the non-polar solvent is evaporated, and the IONPs dispersed in

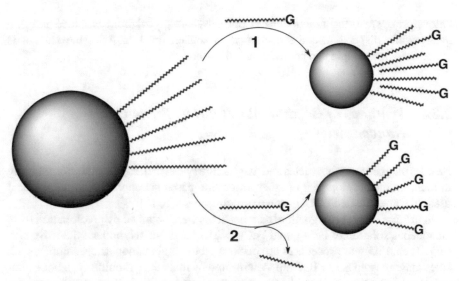

Scheme 2.1 Representation of the main experimental approaches used for surface functionalization. **1** Surfactant addition and **2** surfactant exchange according to Ref. [12]

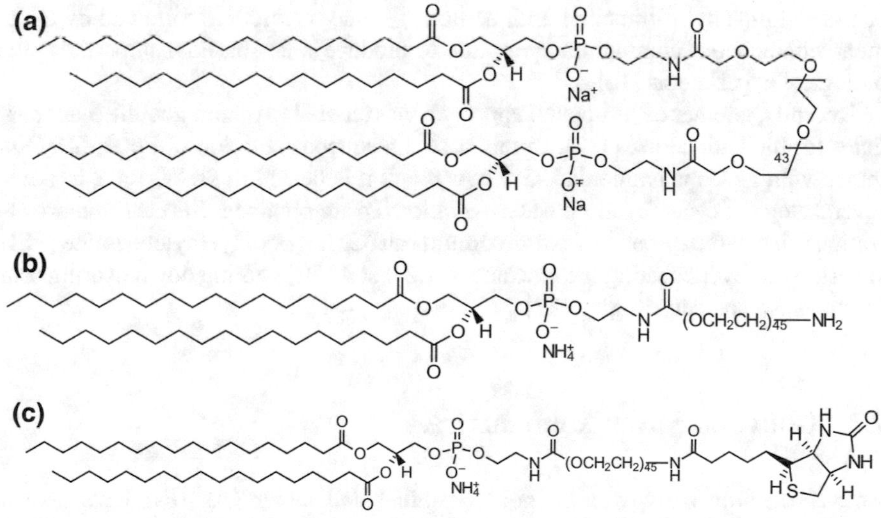

Fig. 2.3 Chemical structures for **a** DSPE-mPEG-2000, **b** DSPE-PEG-NH₂, and **c** DSPE-PEG(2000)Biotin. ®Avanti Polar Lipids, Inc

chloroform. Then, the functionalized polyethylene glycol lipid is added into the nanoparticle dispersion, and the mixture is shaken for 1 h. After the solvent is evaporated, the sample can be dispersed in water.

2.2.4 Synthesis of Goethite-like Hydrous Ferric Oxide Nanoparticles

Besides magnetite nanoparticles and maghemite (γ-Fe_2O_3), some authors are starting to focus on the development of other iron oxide-based nanoparticles for biomedical uses, like goethite (α-FeOOH, iron oxyhydroxide) [16, 17].

In the first report [16], goethite nanoparticles were obtained by precipitation from an aqueous solution of $FeSO_4$ and $FeCl_3$ with excess NaOH, under exhaustive stirring. Then, different procedures are performed to obtain nanoparticle samples with sizes ranging from 5.1 to 10.7 nm, determined by magneto-granulometry. For example, the addition of oleic acid in a similar manner as for magnetite nanoparticles resulted in a toluene-based fluid, and a powder can be obtained by treatment of the fluid with acetone after toluene evaporation. Further modifications include the use of n-decyldimethyl(β-dimethyl-aminoethoxy)-silane methiodide followed by treatment with aqueous ammonium hydroxide to produce water-based nanoparticles for biological experiments [16].

Recently, another experimental approach was studied to obtain goethite nanoparticles for biomedical uses [17]. In this case, an aqueous solution of $FeCl_3 \cdot 6H_2O$ is mixed with aqueous ammonia (25% w/w), and it is heated to 90 °C for 2 h. After several steps of centrifugation, water-soluble NPs are obtained. Surface functionalization using carboxymethyl dextran sodium salt (CMD) or polyethylene imine (PEI) is performed by incubating the reactants for 2 h at 90 °C, cooling down overnight in an autoclave and centrifuging [17].

2.3 Gold and Silver Nanoparticles

Among metallic nanoparticles, gold (AuNPs) and silver (AgNPs) have shown promising behavior for many biomedical applications [18–22]. For example, AuNPs have been tested as antineoplastic agents and drug delivery vectors [3, 5, 6], retinal neovascularization therapies [23], and nerve injuries treatment [24]. Also, AuNPs are of interest in optical and magnetic resonance imaging [25]. In the case of AgNPs, their antimicrobial properties are prime for medical uses, with AgNPs used in dermal wound healing products [26, 27], and in anti-infective corneal replacements [28], for example.

In the following section, some protocols to obtain biocompatible AuNPs and AgNPs are summarized, based on procedures reported as part of their use in biomedical application. However, there is a wide offer of synthetic protocols for AuNPs and AgNPs, which can be revised in the following references [2, 8, 29, 30].

2.3.1 Synthesis of Biocompatible AuNPs

Various types of AuNPs have been developed for biomedical uses, and extensive studies have been reported in radiotherapy (RT), photothermal therapy (PTT), and radiofrequency-induced hyperthermia (RFHT) [21]. So far, there is one FDA-approved nanogold formulation, AuroLase® (Nanospectra Biosciences) consisting in PEGylated silica-gold nanoshells for near-infrared light facilitated thermal ablation, and has been evaluated for PTT in head, neck, and lung tumors [3, 21].

Besides the synthetic protocol employing the reduction of Au^{3+} by citrate at boiling temperature [2] to obtain nanospheres, other reported methodologies produced biocompatible AuNPs, such as nanocages or nanospikes, were obtained by means of a galvanic replacement reaction [31–33]. Post-synthesis modifications to obtain surface functionalization are commonly performed and involve polymer-ligands or other capping agents [18, 29].

AuNPs (Fig. 2.4) used for computed tomography image contrast and radiosensitization [34] are obtained as follows. First, 1.9 nm dodecanethiol-capped AuNP is obtained according to the Brust–Schiffrin method [35] using a water-toluene reduction of $HAuCl_4$ by $NaBH_4$ in the presence of the alkanethiol and employing tetraoctylammonium bromide as the phase transfer agent [34]. Afterward, the AuNPs dissolved in toluene are combined with a solution of the amphiphilic diblock copolymer polyethylene oxide (4K)-polycaprolactone (3K) (PEG-b-PGL) and added to a vessel containing water to produce an emulsion. After purification using differential centrifugation, gold-loaded polymeric micelles of six different sizes (25–150 nm) are achieved as final product (Fig. 2.4).

Another strategy to obtain AuNPs for RT applications at a reduced dose consists of the post-synthesis modification of citrate-capped AuNPs at pH 8.5 with a SH-$PEG2k$-OCH_3 and lipoic acid mixture [36]. The modified AuNPs (Fig. 2.5) aggregate within tumors due to the decreased pH in endosomes and lysosomes, becoming NIR-absorbent and enabling tumor-specific heating upon NIR illumination.

AuNPs approximately 9 nm in size with potential for peripheral nerve regeneration have been synthesized using a green approach [24, 37]. Concisely, a solution of $HAuCl_4$ is mixed with an ethanolic extract form the matured leaves of *Centella asiatica* and constantly stirred at room temperature for 2.5 h until a stable ruby-red color is observed [37]. The high phenolic content in the extract is responsible for the reduction of Au^{3+} to Au^0.

Biocompatible gold nanorods (AuNRs) for PTT can also be obtained using a seed-mediated template-assisted protocol [38, 39]. First, AuNPs seeds are produced by the reduction of $HAuCl_4$ with ice-cold $NaBH_4$ in the presence of cetyltrimethyl

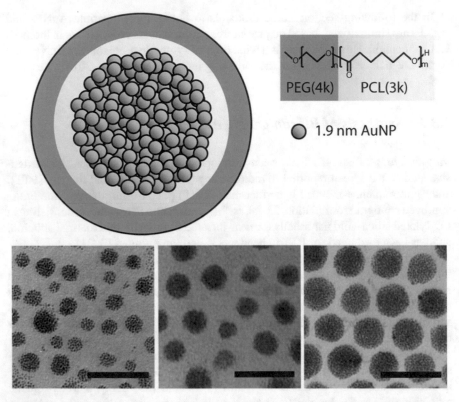

Fig. 2.4 Schematic representation (top) and TEM images corresponding to three different sizes (bottom) of gold-loaded polymeric micelles useful for radiation therapy (scale bars = 100 nm). Adapted with permission from Ref. [34]. Copyright © 2014 American Chemical Society

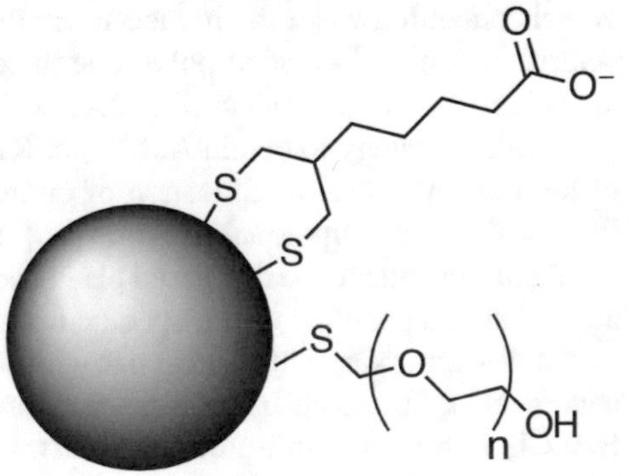

Fig. 2.5 Schematic representation of PEG and lipoic acid-coated AuNP. Adapted with permission from Ref. [36]. Copyright® 2014 Elsevier

ammonium bromide (CTAB). After, 250 μL of the seed solution is added to a growth solution containing CTAB, AgNO₃, HAuCl₄, ascorbic acid, and sulfuric acid and incubated at 30 °C for 20 h. Then, the AuNRs (aspect ratio = 3) are centrifuged and mixed with a solution of SH-PEG-COOH for 2 h. Finally, arginine-glycine-aspartate (RGD) peptides are covalently attached to the PEG-AuNRs via amide bonds [38].

2.3.2 Synthesis of Biocompatible AgNPs

The chemical reduction of silver ions by NaBH₄ in the presence of citrate anion as the stabilizing agent is a common methodology to produce AgNPs in aqueous media [26, 27]. Variations in the protocol can include:

- Different concentration of reactants
- Order of reactant addition
- Reaction time.

For example, AgNPs synthesized using a final ratio of Ag^+:NaBH₄:citrate (1:5:3) [40] were successfully employed in wound healing experiments with similar efficacy compared to a commercial AgNP grafted dressing [27]. In another approach, sodium borohydride in powder is added to the solution containing Ag^+ and citrate to reach a reactants ratio 2:1:7 in that given order and allowed to react overnight under magnetic stirring [26].

An alternative synthetic strategy for biocompatible AgNPs is based on a photochemical methodology using Irgacure-2959® as a photo-initiator [41–43]. Successful preparation and incorporation of AgNPs within collagen hydrogels for corneal implants were achieved recently [28]. Briefly, AgNPs are synthesized under nitrogen by UVA irradiation of silver nitrate in the presence of Irgacure-2959® and citrate during 30–40 min at room temperature. Decreasing the amount of citrate three times leads to large spherical particles. Then, the obtained AgNPs are reshaped at room temperature using LED illumination (590 or 740 nm) for up to 144 h. Post-stabilization of these nanomaterials is assessed using the LL37-SH peptide, and they can be incorporated in collagen hydrogels [28]. The protocol is summarized in Scheme 2.2.

2.4 Graphene Oxide and Carbon Dots

Allotropes of carbon at the nanoscale such as carbon nanotubes, graphene oxide sheets, and carbon dots are also receiving great attention for biomedical purposes [2, 4, 44, 45]. They exhibit many unique physical and chemical properties, such as high conductivity. In the following section, we present some protocols employed to obtain graphene oxide and carbon dots with potential in medical applications.

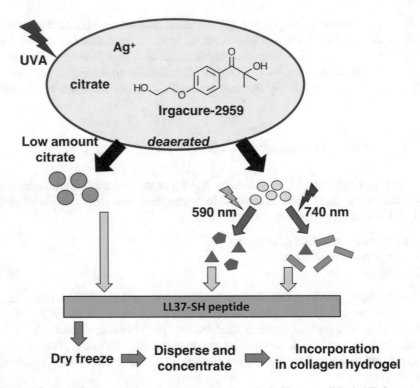

Scheme 2.2 Flow diagram of the protocol proposed to obtain biocompatible AgNP for corneal implants according to Ref. [28]

2.4.1 Graphene Oxide

Graphene oxide (GO) can be obtained through the Hummers method [46]. In a typical procedure, graphite is mixed with concentrated sulfuric acid and vigorously stirred for half an hour before adding sodium nitrate. The mixture is cooled down to 0 °C and stirred for 2 h. Then, potassium permanganate is added over a period of 1 h. After increasing the temperature to 35 °C for 2 h, the reaction is quenched by the addition of ice water and hydrogen peroxide. Subsequently, the GO is washed and dried at 40 °C for 24 h [47]. Reduced graphene can be produced by reduction of GO using ascorbic acid at room temperature. Using this approach, obtained GO sheets were tested in stem cells differentiation [47] and cardiac repair applications [48, 49].

Aiming to apply the GO for gene delivery therapies, a polyamidoamine (PAMAM) conjugated GO has been proposed (Fig. 2.6) [50]. The GO is synthesized using the modified Hummers method through replacing sodium nitrate by phosphoric acid. Also, the temperature is held at 0 °C for 48 h before quenching. To synthesize the PAMAM conjugate, a previous step involves functionalizing the GO with azide groups, and finally, the PAMAM conjugated GO is obtained through a "click" chemistry reaction [50].

(a)

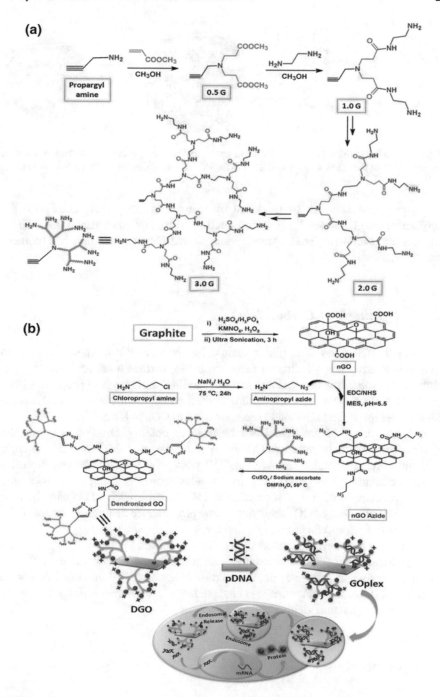

(b)

Fig. 2.6 Schematic diagram of the main steps involved in the synthesis of **a** PAMAM, **b** PAMAM conjugated GO and gene delivery. Reprinted with permission from Ref. [50]. Copyright © 2015 Royal Society of Chemistry

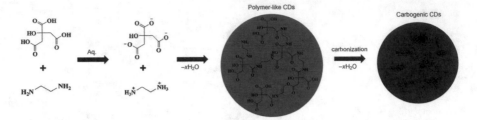

Fig. 2.7 Synthetic route to obtain CDs from citric acid and ethylene diamine. Reprinted with permission from Ref. [55]. Copyright© 2013 Wiley-VCH Verlag GmbH & Co. KGaA, Weinheim

3D graphene inks can be produced by using a biocompatible elastomer, the polylactide-co-glycolide (PLG), in conjunction with graphene flakes (60%). These inks perform excellently well for fabricating 3D scaffolds for electronic and biomedical applications [51].

2.4.2 Synthesis of Carbon Dots

Carbon dots (CDs) have emerged as candidates for biomedical applications due to their unique advantages, including facile synthesis, surface functionalization capability, water solubility, low toxicity, and outstanding photophysical properties [44]. Different synthetic approaches, including hydrothermal, microwave, thermal decomposition, template, and ultrasound methods, lead to CDs [44, 52–54].

A large variety of precursors can be used to obtain CDs. For example, the hydrothermal treatment produces CDs in an autoclave, at 200 °C for 5 h with citric acid and ethylene diamine (EDA) (Fig. 2.7) with good yield (58%) and high fluorescent quantum yield (80%) [55]. In a similar approach, using methionine and EDA as precursors, after a 10 h reaction at 250 °C, sulfur-rich CDs were obtained [56]. These sulfur-doped CDs show interesting properties to distinguish normal from bone-related disease cells in imaging diagnosis.

Recently, a green synthetic strategy has been proposed to obtain PEG-passivated CDs for drug delivery purposes. Briefly, gelatin is diluted in water, and then, PEG is added under stirring. After 30 min, the mixture is heated for 10 min in a microwave oven (600 W). Once a color change is observed, the solution is centrifuged to purify the CDs-PEG (quantum yield: 34%) [57].

2.5 Nanoceria

Cerium oxide nanoparticles (CeO_2) have attracted great interest in nanomedicine due to their anti-inflammatory, antioxidant, and anti-bactericidal activities, as well as their

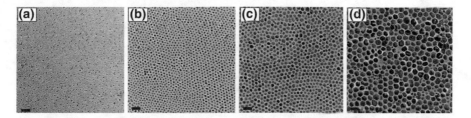

Fig. 2.8 TEM images of nanoceria of several average sizes obtained through thermal decomposition of different precursors. **a** [2.9 nm]; Ce(acac)$_3$, 1,2-hexadecanediol, and oleylamine (1:5:3) at 315 °C for 2 h. **b** [3.6 nm]; Ce(NO$_3$)$_3$·6H$_2$O and oleylamine (1:3). **c** [6.2 nm]; Ce(NO$_3$)$_3$·6H$_2$O and oleylamine (10:30). **d** [9.7 nm]; Ce(NO$_3$)$_3$·6H$_2$O, oleylamine, octadecylamine, and water (1:1.5:1.5:4) at 260 °C for 2 h. Reprinted with permission from Ref. [59]. Copyright© 2012 American Chemical Society

potential angiogenic function. Application of this material alone or combination with others has been reported to heal various tissues (bone, skin, cardiac, and nerve) [58].

Methodologies to obtain nanoceria include chemical and green-based methods. Co-precipitation, sonochemical, hydrothermal, solvothermal, sol-gel, and microwave are the main chemical synthetic strategies employed, while the "green" methods are based on the use of plants, nutrients, and biopolymers [58]. In this section, we present a few protocols for obtaining nanoceria.

Thermal decomposition of cerium precursors (including Ce(acac)$_3$, cerium (III) oleylamine, Ce(NO$_3$)$_3$, and Ce(OH)$_4$) at high temperatures (>200 °C) in 1-octadecene produces several ceria nanocrystals with tunable sizes and shapes. Transference of nanoceria from the organic solvent to water can be achieved using amphiphilic polymers and oleic acid [59]. The resulting nanoparticles are 3–10 nm in diameter depending on the reaction conditions (Fig. 2.8). In a typical procedure, the reaction begins with the dissolution of the cerium precursor in the selected organic solvent. For example, Ce(acac)$_3$ is mixed with 1,2-hexadecanediol and oleylamine in 1-octadecene at room temperature. The mixture is stirred at 80 °C for 30 min and then heated to 315 °C for 2 h [59].

As mentioned above, the synthesis of CeO$_2$ nanoparticles can also be attained using green approaches. For example, using different plants such as *Gloriosa superba*, *Acalipha indica*, and *Aloe vera* nanoceria particles of 5, 36, and ≈64 nm were obtained, respectively. Generally, the plant extract is mixed with CeCl$_3$ at 80 °C for 2–4 h and then heated up to 400 °C for 2 h [60]. Another strategy uses *Rubia cordifolia* leaves extract mixed with Ce(NO$_3$)$_3$·6H$_2$O at 120 °C for 4–6 h before calcinating the formed precipitate at 500 °C for 4 h to obtain 22 nm CeO$_2$ nanoparticles [61]. Other sustainable methods involve the use of nutrients like honey, egg white proteins, and biopolymers [60].

Functionalization of nanoceria has also been attempted for biomedical applications [62–65]. For instance, polyacrylic acid (PAA)-coated cerium oxide nanoparticles were conjugated with folic acid using a "click" chemistry strategy (Fig. 2.9) to use in combination therapy for lung cancer [62].

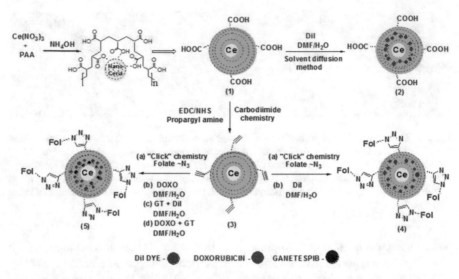

Fig. 2.9 Scheme of the reaction protocol to obtain functionalized nanoceria for treatment of lung cancer. Reprinted with permission from Ref. [62]. Copyright© 2017 American Chemical Society

2.6 Characterization of Nanomaterials

Routine analysis of nanomaterials for biomedical applications requires an excellent characterization of physical, chemical, and biological properties. In this section, we present in Table 2.1, a summary of the main techniques employed to achieve this goal, grouped by method classification, and included key information obtained. We also point to some examples of their application that were tested for nanomedicine uses.

Analytical techniques for nanomaterial characterization can be viewed as a battery of tools necessary to assure reproducibility or robustness when replicating to the same material in another assay. Briefly, microscopy techniques include those based on electron beams as the "illumination" source (transmission electron microscopy (TEM) and scanning electron microscopy (SEM)) and those more traditionally employed in biology, such as confocal microscopy. TEM and SEM can be coupled to spectroscopical methods such as energy-dispersive X-ray spectroscopy (EDX, EDAX or EDS) or X-ray diffraction (SAED).

Those techniques classified as spectroscopical contain the common ones used in organic chemistry such as UV-visible and near-infrared (NIR) absorption, Fourier-transform infrared spectroscopy (FT-IR), nuclear magnetic resonance (NMR), fluorescence, and others more specific like Raman, atomic absorption (AAS) or inductively coupled atomic spectroscopy (ICP-OES), and X-ray photoelectron spectroscopy (XPS).

Table 2.1 Summary of analytical techniques commonly used for characterizations of nanomaterials in biomedical literature

Method classification	Technique	Information	Application examples for NPs in nanomedicine	References
Microscopy	TEM/high resolution TEM (HR-TEM)	Size, polydispersity, and morphology	AuNPs, SPION, nanoceria, GO, and CD	[31–34, 36, 38, 47, 48, 50, 57, 61, 63, 65, 68–74]
	Bio-TEM	Fluorophore labeling	AuNPs	[31]
	SEM/field emission-SEM (FE-SEM)	Size, morphology	AuNPs, nanoceria, and GO	[36, 47, 61, 73]
	Atomic force microscopy (AFM)	Height, topography	GO	[48, 50]
	Confocal laser scanning microscopy (CLSM)	Cell viability	AuNPs	[31]
Spectroscopy	Energy-dispersive X-ray spectroscopy (EDX or EDAX)	Chemical composition	Nanoceria	[61]
	UV-visible-(NIR) spectroscopy	Surface plasmon	AuNPs, IONP, CDs, and nanoceria	[31–34, 36, 38, 57, 61, 72, 74]
	Fourier-transform infrared spectroscopy (FT-IR) Attenuated total reflectance (ATR-IR)	Surface coating characterization and chemical composition	AuNPs, IONP, nanoceria, GO, and CDs	[31, 47, 48, 50, 57, 61, 69, 72]
	Nuclear magnetic resonance (^{1}H NMR and ^{13}C NMR)	Surface coating characterization and chemical composition	GO	[50]
	Raman spectroscopy	Surface coating characterization and chemical composition	GO	[48, 50]

(continued)

Table 2.1 (continued)

Method classification	Technique	Information	Application examples for NPs in nanomedicine	References
	Microwave plasma-atomic emission spectrometry (MP-AES)	Quantification	IONP	[69]
	Inductively coupled plasma-optical emission spectroscopy (ICP-OES)	Quantification	SPION and nanoceria	[68, 73]
	Atomic absorption spectroscopy (AAS)	Quantification	AuNPs	[31]
	Fluorescence spectroscopy	Fluorophores labeling and photoluminescence	AuNPs, IONP, and CDs	[31, 57, 72]
	X-ray photoelectron spectroscopy (XPS)	Surface coated characterization and chemical composition	AuNPs, SPION, nanoceria, and GO	[31, 61, 65, 68]
Cells assays	MTT assay	Cytotoxicity and cell viability	AuNPs, SPION, nanoceria, GO, and CD	[31, 32, 50, 57, 61, 63, 65, 68, 70, 72]
	WST-1 assay	Cell proliferation and cytotoxicity	IONP and nanoceria	[69, 73]
	WST-8 or CCK-8	Cell proliferation and cytotoxicity	IONP and GO	[48, 71]
	Quant-iT^{TM} PicoGreen® dsDNA assay	Cell proliferation	Nanoceria	[73]
	Hemolytic assay	Hemocompatibility	IONP	[71]
	Annexin V-FITC	Apoptosis and cytotoxicity	IONP	[72]
	ApoTox-Glo^{TM} triplex assay	Apoptosis, cytotoxicity, and cell viability	Nanoceria	[65]

(continued)

Table 2.1 (continued)

Method classification	Technique	Information	Application examples for NPs in nanomedicine	References
Other methods	Selected area electron diffraction (SAED)	Crystallinity	Nanoceria and GO	[48, 61]
	ICP-mass spectrometry (ICP-MS)	Quantification	AuNPs, nanoceria, and GO	[33, 48, 63]
	X-ray diffraction (XRD)	Crystallinity and chemical composition	IONP, nanoceria, and GO	[57, 61, 65, 68]
	Thermogravimetric analysis (TGA)	Surface coating quantification	IONP, nanoceria, and GO	[47, 65, 71]
	Superconducting quantum interference device (SQUID) magnetometry	Magnetic properties	SPION	[68, 70]
	Relaxometry	Relaxation variables	IONP	[71, 72]
	Dynamics light scattering (DLS)	Hydrodynamic size and polydispersity	AuNPs, IONP, nanoceria, GO, and CD	[31, 33, 34, 36, 50, 57, 63, 65, 68–73]
	Zeta potential	Surface charge and flocculation tendency	AuNPs, SPION, nanoceria, GO, and CD	[31, 34, 50, 57, 63, 65, 68, 69, 72–74]
	Total antioxidant capacity assay	Anti-oxidant substances concentration and efficacy	Nanoceria	[73]

Besides, there are assays commonly performed in cells studies to determine cytotoxicity and other aspects related to biocompatibility. These assays contain the colorimetric ones using 3-(4,5-dimethylthiazol-2-yl)-2,5-diphenyl tetrazolium bromide (MTT), [2-(4-Iodophenyl)-3-(4-nitrophenyl)-5-(2,4-disulfophenyl)-2H-tetrazolium] (WST-1), or another tetrazolium salt (WST-8 or CCK-8). Also, assays based on fluorescence are performed such as the use of annexins proteins conjugated with a fluorophore, for instance, annexin V-fluorescein isothiocyanate (FITC).

Other techniques commonly used for nanomaterials characterization are included in Table 2.1.

2.7   Concluding Remarks

Numerous nanomaterials have been extensively studied in recent years for diverse biomedical applications, ranging from drug delivery to imaging techniques. There are several reports that demonstrate that these materials were successfully used in both "in vitro" and "in vivo" assays.

Although NM can be obtained by either a *Top-down* or *Bottom-up* methodology, throughout this chapter, we have reviewed those synthesized using the *Bottom-up* method. These strategies are more versatile and permit the production of a wide variety of nanomaterials starting from simple reactants, such as inorganic salts.

Synthetic approaches must be consistent with the intended application of the material so, for biomedical purposes, "biocompatibility" is mainly sought. The main limitations of these synthetic methodologies, usually, are due to the fact that the nanomaterial is obtained in a non-polar organic solvent; and for medical uses, it needs to be transferred to water or transformed in a water-soluble compound. This can be achieved by post-synthesis surface modification of the NMs. Also, the selection of an appropriate synthetic route will depend on the desired nanomaterial features (e.g., size, shape, biological activity, etc.). Moreover, batch-to-batch reproducibility is a key factor, and in our opinion, the optimization of synthetic procedures using design of experiment methods can be a good solution to this problem [66].

Regarding the characterization tools used for NMs, they have to be selected according to the information requested (Table 2.1), and it is always required to analyze the nanomaterial using a set of analytical techniques, so the nanomaterial product is accompanied by a prospect with different parameters specifications. This essential requisite has made the characterization stage time-consuming and expensive [67].

We expect this chapter provides the readers with a helpful overview of the synthetic procedures available for different nanomaterials with biomedical potential.

Acknowledgements   We would like to thank to all the researchers whose cited work has made possible this chapter. Our deep gratitude to the CONICET and SECyT-UNC for financial support. N.L.P. is a research member of the Consejo Nacional de Investigaciones Científicas y Técnicas (CONICET) of Argentine. R.N.N. and M.A.M.T. are grateful recipients of graduated fellowships from CONICET.

Disclosure   All authors have read and approved the final version.

References

1. Wagner V, Dullaart A, Bock A-K, Zweck A. The emerging nanomedicine landscape. Nat Biotechnol. 2006;24(10):1211–7.
2. Aguilar ZP. Nanomaterials for medical applications. Waltham, USA: Elsevier Ltd; 2013. p. 461.
3. Anselmo AC, Mitragotri S. Nanoparticles in the clinic. Bioeng Transl Med. 2016;1(1):10–29.
4. Mclaughlin S, Podrebarac J, Ruel M, Suuronen E, McNeill B, Alarcon E. Nano-engineered biomaterials for tissue regeneration: what has been achieved so far? Front Mater. 2016;3:67435.

5. Bobo D, Robinson KJ, Islam J, Thurecht KJ, Corrie SR. Nanoparticle-based medicines: a review of FDA-approved materials and clinical trials to date. Pharm Res. 2016;33(10):2373–87.
6. Ventola CL. Progress in nanomedicine: approved and investigational nanodrugs. PT. 2017;42(12):742–55.
7. Hornyak GL, Dutta J, Tibbals HF, Rao AK. Introduction to nanoscience. Boca Raton: CRC Press; 2008.
8. Pacioni NL, Borsarelli CD, Rey V, Veglia AV. Synthetic routes for the preparation of silver nanoparticles. A mechanistic perspective. In: Alarcón EI, Griffith M, Udekwu KI, editors. Silver nanoparticles applications. Switzerland: Springer; 2015. p. 13–46.
9. Hu Y, Mignani S, Majoral J-P, Shen M, Shi X. Construction of iron oxide nanoparticle-based hybrid platforms for tumor imaging and therapy. Chem Soc Rev. 2018;47(5):1874–900.
10. Song C, Sun W, Xiao Y, Shi X. Ultrasmall iron oxide nanoparticles: synthesis, surface modification, assembly, and biomedical applications. Drug Discov Today. 2019;24(3):835–44.
11. Yu WW, Falkner JC, Yavuz CT, Colvin VL. Synthesis of monodisperse iron oxide nanocrystals by thermal decomposition of iron carboxylate salts. Chem Commun. 2004;20:2306.
12. Xie J, Peng S, Brower N, Pourmand N, Wang SX, Sun S. One-pot synthesis of monodisperse iron oxide nanoparticles for potential biomedical applications. Pure Appl Chem. 2006;78(5):1003–14.
13. Han J, Kim B, Shin J-Y, Ryu S, Noh M, Woo J, Park J-S, Lee Y, Lee N, Hyeon T, Choi D. Iron oxide nanoparticle-mediated development of cellular gap junction crosstalk to improve mesenchymal stem cells' therapeutic efficacy for myocardial infarction. ACS Nano. 2015;9(3):2805–19.
14. Xu H, Aguilar ZP, Yang L, Kuang M, Duan H, Xiong Y, Wei H, Wang A. Antibody conjugated magnetic iron oxide nanoparticles for cancer cell separation in fresh whole blood. Biomaterials. 2011;32(36):9758–65.
15. Gao X, Cui Y, Levenson RM, Chung LWK, Nie S. In vivo cancer targeting and imaging with semiconductor quantum dots. Nat Biotechnol. 2004;22(8):969–76.
16. Segal I, Zablotskaya A, Lukevics E, Maiorov M, Zablotsky D, Blums E, Mishnev A, Georgieva R, Shestakova I, Gulbe A. Preparation and cytotoxic properties of goethite-based nanoparticles covered with decyldimethyl(dimethylaminoethoxy) silane methiodide. Appl Organomet Chem. 2010;24(3):193–7.
17. Lunin AV, Kolychev EL, Mochalova EN, Cherkasov VR, Nikitin MP. Synthesis of highly-specific stable nanocrystalline goethite-like hydrous ferric oxide nanoparticles for biomedical applications by simple precipitation method. J Colloid Interface Sci. 2019;541:143–9.
18. Cobley C, Chen J, Cho E, Wang L, Xia Y. Gold nanostructures: a class of multifunctional materials for biomedical applications. Chem Soc Rev. 2011;40(1):44.
19. Alarcón EI, Griffith M, Udekwu K, editors. Silver nanoparticle applications. In the fabrication and design of medical and biosensing devices. Switzerland: Springer International Publishing; 2015.
20. Hosoyama K, Ahumada M, McTiernan CD, Bejjani J, Variola F, Ruel M, Xu B, Liang W, Suuronen EJ, Alarcon EI. Multi-functional thermo-crosslinkable collagen-metal nanoparticle composites for tissue regeneration: nanosilver vs. nanogold. RSC Adv. 2017;7(75):47704–8.
21. Beik J, Khateri M, Khosravi Z, Kamrava SK, Kooranifar S, Ghaznavi H, Shakeri-Zadeh A. Gold nanoparticles in combinatorial cancer therapy strategies. Coord Chem Rev. 2019;387:299–324.
22. Duncan B, Kim C, Rotello VM. Gold nanoparticle platforms as drug and biomacromolecule delivery systems. J Control Release. 2010;148(1):122–7.
23. Kim JH, Kim MH, Jo DH, Yu YS, Lee TG, Kim JH. The inhibition of retinal neovascularization by gold nanoparticles via suppression of VEGFR-2 activation. Biomaterials. 2011;32(7):1865–71.
24. Das S, Sharma M, Saharia D, Sarma KK, Sarma MG, Borthakur BB, Bora U. In vivo studies of silk based gold nano-composite conduits for functional peripheral nerve regeneration. Biomaterials. 2015;62:66–75.
25. Gitanjali A, Brahmkhatri VP, Atreya HS. Nanomaterial based magnetic resonance imaging of cancer. J Indian Inst Sci. 2014;94(4):423–53.

26. Liu X, Lee P-Y, Ho C-M, Lui VCH, Chen Y, Che C-M, Tam PKH, Wong KKY. Silver nanoparticles mediate differential responses in keratinocytes and fibroblasts during skin wound healing. Chem Med Chem. 2010;5(3):468–75.
27. Tian J, Wong KKY, Ho C-M, Lok C-N, Yu W-Y, Che C-M, Chiu J-F, Tam PKH. Topical delivery of silver nanoparticles promotes wound healing. Chem Med Chem. 2007;2(1):129–36.
28. Alarcon E, Vulesevic B, Argawal A, Ross A, Bejjani P, Podrebarac J, Ravichandran R, Phopase J, Suuronen E, Griffith M. Coloured cornea replacements with anti-infective properties: expanding the safe use of silver nanoparticles in regenerative medicine. Nanoscale. 2016;8(12):6484–9.
29. Saha K, Agasti S, Kim C, Li X, Rotello V. Gold nanoparticles in chemical and biological sensing. Chem Rev. 2012;112(5):2739–79.
30. Jones M, Osberg K, Macfarlane R, Langille M, Mirkin C. Templated techniques for the synthesis and assembly of plasmonic nanostructures. Chem Rev. 2011;111:3736–827.
31. Ma N, Jiang Y-W, Zhang X, Wu H, Myers JN, Liu P, Jin H, Gu N, He N, Wu F-G, Chen Z. Enhanced radiosensitization of gold nanospikes via hyperthermia in combined cancer radiation and photothermal therapy. ACS Appl Mater Interfaces. 2016;8(42):28480–94.
32. Zhang YS, Wang Y, Wang L, Wang Y, Cai X, Zhang C, Wang LV, Xia Y. Labeling human mesenchymal stem cells with gold nanocages for *in vitro* and *in vivo* tracking by two-photon microscopy and photoacoustic microscopy. Theranostics. 2013;3(8):532–43.
33. Park J, Park J, Ju EJ, Park SS, Choi J, Lee JH, Lee KJ, Shin SH, Ko EJ, Park I, Kim C. Multifunctional hollow gold nanoparticles designed for triple combination therapy and CT imaging. J Control Release. 2015;207:77–85.
34. Al Zaki A, Joh D, Cheng Z, De Barros ALB, Kao G, Dorsey J, Tsourkas A. Gold-loaded polymeric micelles for computed tomography-guided radiation therapy treatment and radiosensitization. ACS Nano. 2014;8(1):104–12.
35. Brust M, Walker M, Bethell D, Schiffrin DJ, Whyman R. Synthesis of thiol-derivatised gold nanoparticles in a two-phase liquid–liquid system. Chem Commun. 1994;7:801–2.
36. Hainfeld JF, Lin L, Slatkin DN, Avraham Dilmanian F, Vadas TM, Smilowitz HM. Gold nanoparticle hyperthermia reduces radiotherapy dose. Nanomedicine. 2014;10(8):1609–17.
37. Das RK, Borthakur BB, Bora U. Green synthesis of gold nanoparticles using ethanolic leaf extract of *Centella asiatica*. Mater Lett. 2010;64(13):1445–7.
38. Li P, Shi Y-W, Li B-X, Xu W-C, Shi Z-L, Zhou C, Fu S. Photo-thermal effect enhances the efficiency of radiotherapy using Arg-Gly-Asp peptides-conjugated gold nanorods that target αvβ3 in melanoma cancer cells. J Nanobiotechnol. 2015;13(52).
39. Pan B, Ao L, Gao F, Tian H, He R, Cui D. End-to-end self-assembly and colorimetric characterization of gold nanorods and nanospheres via oligonucleotide hybridization. Nanotechnology. 2005;16(9):1776–80.
40. Jin R, Cao Y, Mirkin CA, Kelly KL, Schatz GC, Zheng JG. Photoinduced conversion of silver nanospheres to nanoprisms. Science. 2001;294(5548):1901–3.
41. Alarcon E, Udekwu K, Skog M, Pacioni N, Stamplecoskie K, González-Béjar M, Polisetti N, Wickham A, Richter-Dahlfors A, Griffith M, Scaiano JC. The biocompatibility and antibacterial properties of collagen-stabilized, photochemically prepared silver nanoparticles. Biomaterials. 2012;33(19):4947–56.
42. Scaiano J, Netto-Ferreira J, Alarcon E, Billone P, Alejo C, Crites C, Decan M, Fasciani C, González-Béjar M, Hallett-Tapley G, Grenier M, McGilvray K, Pacioni N, Pardoe A, René-Boisneuf L, Schwartz-Narbonne R, Silvero M, Stamplecoskie K, Wee T. Tuning plasmon transitions and their applications in organic photochemistry. Pure Appl Chem. 2011;83(4):913–30.
43. Stamplecoskie K, Scaiano J. Light emitting diode irradiation can control the morphology and optical properties of silver nanoparticles. J Am Chem Soc. 2010;132(6):1825–7.
44. Ghosal K, Ghosh A. Carbon dots: the next generation platform for biomedical applications. Mater Sci Eng C. 2019;96:887–903.
45. Tran PA, Zhang L, Webster TJ. Carbon nanofibers and carbon nanotubes in regenerative medicine. Adv Drug Deliv Rev. 2009;61(12):1097–114.
46. Hummers WS, Offeman RE. Preparation of graphitic oxide. J Am Chem Soc. 1958;80(6):1339–1339.

47. Yang D, Li T, Xu M, Gao F, Yang J, Yang Z, Le W. Graphene oxide promotes the differentiation of mouse embryonic stem cells to dopamine neurons. Nanomedicine. 2014;9(16):2445–55.
48. Park J, Kim B, Han J, Oh J, Park S, Ryu S, Jung S, Shin J-Y, Lee BS, Hong BH, Choi D. Graphene oxide flakes as a cellular adhesive: prevention of reactive oxygen species mediated death of implanted cells for cardiac repair. ACS Nano. 2015;9(5):4987–99.
49. Park J, Kim YS, Ryu S, Kang WS, Park S, Han J, Jeong HC, Hong BH, Ahn Y, Kim B-S. Graphene potentiates the myocardial repair efficacy of mesenchymal stem cells by stimulating the expression of angiogenic growth factors and gap junction protein. Adv Funct Mater. 2015;25(17):2590–600.
50. Sarkar K, Madras G, Chatterjee K. Dendron conjugation to graphene oxide using click chemistry for efficient gene delivery. RSC Adv. 2015;5(62):50196–211.
51. Jakus AE, Secor EB, Rutz AL, Jordan SW, Hersam MC, Shah RN. Three-dimensional printing of high-content graphene scaffolds for electronic and biomedical applications. ACS Nano. 2015;9(4):4636–48.
52. Jelinek R. Carbon-dot synthesis. In: Carbon quantum dots. Cham: Springer International Publishing; 2017. p. 5–27.
53. Xu Q, Kuang T, Liu Y, Cai L, Peng X, Sreenivasan Sreeprasad T, Zhao P, Yu Z, Li N. Heteroatom-doped carbon dots: synthesis, characterization, properties, photoluminescence mechanism and biological applications. J Mater Chem B. 2016;4(45):7204–19.
54. Zhang J, Yu S-H. Carbon dots: large-scale synthesis, sensing and bioimaging. Mater Today. 2016;19(7):382–93.
55. Zhu S, Meng Q, Wang L, Zhang J, Song Y, Jin H, Zhang K, Sun H, Wang H, Yang B. Highly photoluminescent carbon dots for multicolor patterning, sensors, and bioimaging. Angew Chem. 2013;125(14):4045–9.
56. Zhu P, Lyu D, Shen PK, Wang X. Sulfur-rich carbon dots as a novel fluorescent imaging probe for distinguishing the pathological changes of mouse bone cells. J Lumin. 2019;207:620–5.
57. Arsalani N, Nezhad-Mokhtari P, Jabbari E. Microwave-assisted and one-step synthesis of PEG passivated fluorescent carbon dots from gelatin as an efficient nanocarrier for methotrexate delivery. Artif Cells Nanomed Biotechnol. 2019;47(1):540–7.
58. Kargozar S, Baino F, Hoseini SJ, Hamzehlou S, Darroudi M, Verdi J, Hasanzadeh L, Kim H-W, Mozafari M. Biomedical applications of nanoceria: new roles for an old player. Nanomedicine. 2018;13(23):3051–69.
59. Lee SS, Zhu H, Contreras EQ, Prakash A, Puppala HL, Colvin VL. High temperature decomposition of cerium precursors to form ceria nanocrystal libraries for biological applications. Chem Mater. 2012;24(3):424–32.
60. Charbgoo F, Ahmad M, Darroudi M. Cerium oxide nanoparticles: green synthesis and biological applications. Int J Nanomed. 2017;12:1401–13.
61. Sisubalan N, Ramkumar VS, Pugazhendhi A, Karthikeyan C, Indira K, Gopinath K, Hameed ASH, Basha MHG. ROS-mediated cytotoxic activity of ZnO and CeO$_2$ nanoparticles synthesized using the *Rubia cordifolia* L. leaf extract on MG-63 human osteosarcoma cell lines. Environ Sci Pollut Res Int. 2018;25(11):10482–92.
62. Sulthana S, Banerjee T, Kallu J, Vuppala SR, Heckert B, Naz S, Shelby T, Yambem O, Santra S. Combination therapy of NSCLC using Hsp90 inhibitor and doxorubicin carrying functional nanoceria. Mol Pharm. 2017;14(3):875–84.
63. Kwon HJ, Cha M-Y, Kim D, Kim DK, Soh M, Shin K, Hyeon T, Mook-Jung I. Mitochondria-targeting ceria nanoparticles as antioxidants for alzheimer's disease. ACS Nano. 2016;10(2):2860–70.
64. Nethi SK, Nanda HS, Steele TW, Patra CR, Chen G. Functionalized nanoceria exhibit improved angiogenic properties. J Mater Chem B. 2017;5:9371–83.
65. Hijaz M, Das S, Mert I, Gupta A, Al-Wahab Z, Tebbe C, Dar S, Chhina J, Giri S, Munkarah A, Seal S. Folic acid tagged nanoceria as a novel therapeutic agent in ovarian cancer. BMC Cancer 2016;16(1).
66. Núñez R, Veglia A, Pacioni N. Improving reproducibility between batches of silver nanoparticles using an experimental design approach. Microchem J. 2018;141:110–7.

67. Pacioni NL. Metrology for metal nanoparticles. In: Martínez LMT, Kharissova OV, Kharisov BI, editors. Handbook of ecomaterials. Cham: Springer International Publishing; 2019. p. 2327–42.
68. Huang Y, Mao K, Zhang B, Zhao Y. Superparamagnetic iron oxide nanoparticles conjugated with folic acid for dual target-specific drug delivery and MRI in cancer theranostics. Mater Sci Eng C. 2017;70:763–71.
69. Ali AAA, Hsu F-T, Hsieh C-L, Shiau C-Y, Chiang C-H, Wei Z-H, Chen C-Y, Huang H-S. Erlotinib-conjugated iron oxide nanoparticles as a smart cancer-targeted theranostic probe for MRI. Sci Rep. 2016;6(36650).
70. Li Z, Yi PW, Sun Q, Lei H, Li Zhao H, Zhu ZH, Smith SC, Lan MB, Lu GQM. Ultrasmall water-soluble and biocompatible magnetic iron oxide nanoparticles as positive and negative dual contrast agents. Adv Funct Mater. 2012;22(11):2387–93.
71. Ma D, Chen J, Luo Y, Wang H, Shi X. Zwitterion-coated ultrasmall iron oxide nanoparticles for enhanced T1-weighted magnetic resonance imaging applications. J Mater Chem B. 2017;5:7267–73.
72. Yin T, Zhang Q, Wu H, Gao G, Shapter JG, Shen Y, He Q, Huang P, Qi W, Cui D. In vivo high-efficiency targeted photodynamic therapy of ultra-small Fe_3O_4@polymer-NPO/PEG-Glc@Ce6 nanoprobes based on small size effect. NPG Asia Mater. 2017;9(5):e383.
73. Marino A, Tonda-Turo C, De Pasquale D, Ruini F, Genchi G, Nitti S, Cappello V, Gemmi M, Mattoli V, Ciardelli G, Ciofani G. Gelatin/nanoceria nanocomposite fibers as antioxidant scaffolds for neuronal regeneration. Biochim Biophys Acta Gen Subj. 2017;1861(2):386–95.
74. Vieira S, Vial S, Maia FR, Carvalho M, Reis RL, Granja PL, Oliveira JM. Gellan gum-coated gold nanorods: an intracellular nanosystem for bone tissue engineering. RSC Adv. 2015;5(95):77996–8005.

Chapter 3
Advanced Surface Characterization Techniques in Nano- and Biomaterials

Ricardo A. Zamora, Cristián Gutiérrez-Cerón, Jesum Alves Fernandes and Gabriel Abarca

Abstract Although metallic nanoparticles have been applied in various fields of biomedical engineering research for quite some time, generating new biomaterials with improved regenerative capabilities remains the cornerstone in tissue engineering and regenerative medicine. These materials, once implanted in patients, will ultimately be invaded by endogenous cells, which emphasizes the relevance of surface composition as a critical factor in determining the regenerative potency of a given material. In this chapter, we present a brief revision on fundamental concepts and an up-to-date overview for surface characterization of nano-engineered structures.

3.1 Introduction

Metal nanoparticles and functionalized metal nanoparticles, bimetallic nanoparticles containing metals such as copper, nickel, silver, and gold, and non-metallic nanomaterials have been widely used in diagnosis and therapeutics [1, 2]. The use of nanomaterials in biomedical applications explodes the nanometric sizes of such materials that confer unique physical and chemical properties. There are several ways to determine the average size of a particle in solution [3, 4]. When a light beam reaches a solution or colloidal dispersion, part of the incident radiation may be absorbed, a part is scattered, and the rest is transmitted through the solution [5]. The intensity, polarization, and angular distribution of the light scattered by a colloidal dispersion depend on the size and shape of the particles, the interactions among them, and the difference between the refractive indexes of the particles and the medium [6, 7]. Light scattering measurements have gained relevance in determining the size, shape, and interactions between the particles [8, 9].

R. A. Zamora · C. Gutiérrez-Cerón
Facultad de Química y Biología, Universidad de Santiago de Chile, Santiago, Chile

J. A. Fernandes
School of Chemistry, University of Nottingham, University Park, Nottingham NG7 2RD, UK

G. Abarca (✉)
Facultad de Ciencias, Centro de Nanotecnología Aplicada, Universidad Mayor, Santiago, Chile
e-mail: gabriel.abarca@umayor.cl

© Springer Nature Switzerland AG 2019
E. I. Alarcon and M. Ahumada (eds.), *Nanoengineering Materials for Biomedical Uses*,
https://doi.org/10.1007/978-3-030-31261-9_3

Dynamic light scattering (DLS), small angle X-ray scattering (SAXS), and small angle neutron scattering (SANS) are the most used analytical techniques based on scattering at small angles of the incident radiation [10, 11]. Such dispersion, regardless of their nature, produces a dispersion pattern, which contains structural information of the sample [9, 12]. Whether we talk about visible light scattering (DLS) or X-ray/neutrons (SAXS and SANS, respectively), we must consider that these types of radiation have two main ways of interacting with matter: absorption and scattering [13, 14]. The dispersion can occur with or without the loss of energy, which leads to the generation of scattering waves with different wavelength (i.e., Compton scattering, inelastic scattering) when there is energy transfer or equal wavelength (Rayleigh scattering and Thomson scattering, elastic scattering) than the incident beam when there is no energy transfer [15].

Rayleigh scattering (visible light) and Thomson scattering (X-ray and neutrons) are generated when the incident radiation collides with the electrons without transferring energy [16]. This interaction produces coherent waves and an interference phenomenon, which allows these to be recorded by a detector. The diffraction pattern obtained allows the obtaining of structural information of the material from these waves.

3.2   Small-Angle X-ray Scattering (SAXS) and Small-Angle Neutron Scattering (SANS)

The SAXS and SANS are analytical techniques based on Thomson scattering that allows the structural characterization of nanomaterials in the range of 1–100 nm (nano to mesoscale) [17]. There are other scattering techniques for materials less than 1 nm (ultra-small-angle X-ray scattering, USAXS) and greater than 100 nm (wide-angle X-ray scattering, WAXS) [12, 18].

SAXS uses X-ray with wavelengths of 0.1–0.2 nm and allows characterizing the size, shape, and morphology of synthetic and natural polymers, nanoparticles, and biomaterials, while SANS technique uses neutrons with a wavelength of 0.5 nm approximately and is widely used to study the size and structural dynamics of nanomaterials [12, 17]. Figure 3.1 top shows the fundamental setup of a SAXS and SANS experiment.

When the SAXS and SANS experiments are performed, the data that is recorded by a detector. The detector records a scattering vector (q), which is defined by subtraction of the scattering wave vector (k_1) and scattering wave incident (k_0). The q vector presents elastic scattering, and the modulus of the scattered wave k_1 is equal to k_0. The q vector is represented mathematically by the expression:

$$q(\lambda, \theta) = (4\pi/\lambda)\sin(\theta) \tag{3.1}$$

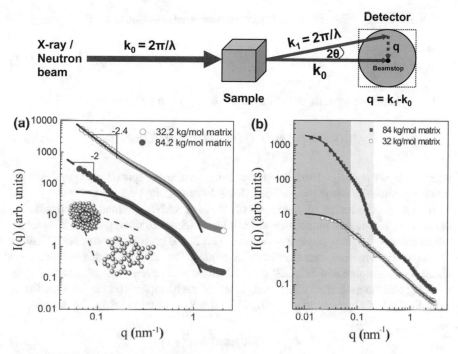

Fig. 3.1 (Top) SAXS and SANS standard setup. (Bottom) PS-grafted Fe_3O_4 nanoparticles measurements profiles, **a** SAXS **b** SANS bottom plots. Reproduced with permission from [19]

The intensity of scattering $I(q)$ is expressed as a function of the scattered vector, which depends on the wavelength (λ) of the incident beam and scattering angle (2θ). The intensity of scattering also is given by (3.2).

$$I(q) = N P(q)S(q) = \Phi \Delta p^2 V P(q)S(q) \tag{3.2}$$

Here, N is the number of particles that scatter the incident beam and the terms $P(q)$ and $S(q)$ represent the form factor and the structure factor, respectively [20]. The form factor is related to the size and shape of the object, and the structure factor is related to the distribution of distances between a particle and its neighbors, being this term an index of the state of aggregation of the sample.

$$P(q) = | \int (\rho(r) - \rho_S)\exp(iqr)dr |^2 \tag{3.3}$$

For a monodisperse system containing spherical particles, the form factor $P(q)$ can be described by (3.4):

$$P(q) = (\rho - \rho_S)^2 V^2 \left(\frac{3J_1(qR)}{qR} \right) \tag{3.4}$$

If we consider spherical particle having core–shell structure,

$$P(q) = \left[(\rho 1 - \rho 2)V 1 \frac{3J_1(qR_1)}{qR_1} + (\rho 2 - \rho S)V(3J1qRqR) \right]^2 \tag{3.5}$$

For an isotropic system, the structure factor $S(q)$ can be described by (3.6),

$$S(q) = 1 + 4\pi n \int (g(r) - 1)(\sin qR)/qRr^2 dr \tag{3.6}$$

where $g(r)$ is the radial distribution function; the above expression depends on the relative position of the particles. For a diluted system, $S(q) \sim 1$.

In practice, the scattering spectrums of SAXS or SANS obtained of the irradiation of samples diluted and monodisperses can be divided into three parts: low-q region, intermediary-q region, and high-q region. The low-q region (the so-called Guinier region) is the region to minimal scattering angles. The analysis of this region allows obtaining for diluted and monodisperse solutions parameters such as radius of gyration (R_g) and the molecular weight (M_w). M_w of the nanoparticles can be obtained by extrapolation of absolute data to $I(0)$, while the R_g is obtained from the expression,

$$I(q) = I(0)\exp\left(R_g^2 q^2/3 \right) \tag{3.7}$$

where the R_g is obtained from the slope $(-R_g^2/3)$ from plot Ln $I(q)$ v/s q^2. The above expression is valid for any particle shapes [18]. When it is known that the shape of the nanoparticles is too elongated or flattened, similar expressions can be used to obtain the corresponding radius [18]. In the intermediary-q region (the so-called Fourier region), the pair distance distribution function can be determined. This phenomenon is possible by an indirect Fourier transformation of the experimental form factor. The calculation of this parameter gives us relevant information about the general shape of the nanoparticle [21].

$$p(r) = 1/2\pi r^2 \int\limits_0^\infty qP(q)\sin(qr)dr \tag{3.8}$$

The high-q region (the so-called Porod region) is the region at large scattering angles. The data from this region is used to establish characteristics of the surface of the nanoparticle, such as the surface-to-volume ratio and specific surface estimation for small particles [21]. For this, it is necessary to obtain the so-called invariant Q. The invariant Q is obtained by integrating the expression $I(q)q^2$.

$$Q = \int\limits_0^\infty I(q)q^2 \tag{3.9}$$

The integration gives the value of invariant Q, which only depends on the volume of the nanoparticle and not on its shape [17]. Additionally, with the invariant Q, it is possible to obtain the Porod volume, which is generally 1.5 or 2 times the M_w of the nanoparticle [22]. Along with this, if the Guinier approximation is used to extrapolate the value of $I(0)$, it is possible to calculate, along with the invariant Q, the M_w of the nanoparticle according to the expression $M_w = I(0)/Q$.

In addition to the Guinier and Porod approximations, there are other methods to analyze samples of monodisperse and diluted nanoparticles. Among the simplest practices described to perform parameter analysis from SAXS and SANS experiments, we can mention the average size determination method (ADM) [23]. This method is exclusively used for the determination of the size distribution and the shape of particles in monodisperse systems and can be used with a model-free approximation or considering the use of one. In more complex cases, in which the particle system is polydispersed, we can mention as methods of analysis the parametric distribution models (PDM), integral transform methods (ITM), and numerical methods (NM) [23].

The parametric distribution models are the most straightforward methods used to determine the size distribution in polydispersed nanoparticle systems. This method assumes a parametric distribution of the size distribution, while the integral transform methods and numerical methods assume a specific original form. Because the SAXS and SANS techniques are highly specific in determining the distribution of sizes and shapes of different types of nanoparticles, their use as characterization techniques has increased sharply in recent years in scientific journals that have as scope the application of nanoparticles in the biomedical area [24, 25].

3.3 SAXS, SANS, and Its Application in Biomedicine

3.3.1 Tailored Nanoparticles with Antiviral and Antibacterial Activities

The use of nanoparticles as therapeutic agents for the control of various viruses and pathogenic bacteria has taken high relevance in recent years due to the wide variety of materials that can be used for their synthesis [26, 27]. Besides, most nanoparticles are easy to synthesize and present a high ductility in the modification of surface physicochemical properties, along with not being cytotoxic against normal mammalian cells [28]. Currently, many reports show the use of functionalized nanoparticles as antiviral and antibacterial properties [29, 30]. Nanostructures with antiviral activity must have a specific size to be able to reversibly interfere with the virus–cell receptor interaction (virustatic nanoparticles) [31]. Another type of antiviral nanoparticles is the so-called virucidal nanoparticles, which acts irreversibly deactivating the action of the virus [31]. One of the nanoparticles used for this purpose is the mesoporous silica nanoparticles ($mSiO_2$). Silva et al. established for the first time that the $mSiO_2$

nanoparticles have virucidal antiviral activity [32]. In their work, they synthesized and characterized $mSiO_2$ nanoparticles by SAXS, which functionalized with diverse chemical groups on the surface to generate virucidal nanoparticles [33]. The use of SAXS allowed establishing that the modification of the surface with different types of substituents did not alter the uniformity of shape of the functionalized $mSiO_2$ nanoparticles compared to non-functionalized ones (both presented a spherical shape with a similar size distribution). Similarly, Sokolowski et al, complementing measures of SAXS and SANS, characterized anionic and cationic SiO_2 nanoparticles [33]. Yi et al. through time-resolved SAXS could establish the mechanism of growth and the kinetics of formation of this $mSiO_2$ nanoparticles, showing how the use of this technique allowed characterizing structural changes during an increase in this type of nanoparticles at the nanometer scale [34].

Other nanoparticles that have shown efficient antiviral activity are silver nanoparticles (AgNPs). Wuithschick et al. using in situ SAXS could synthesize silver nanoparticles and characterize the average radius of these. The use of SAXS allowed them to standardize a method of synthesis of AgNPs with control of the size of these without the use of stabilizers [35].

In the case of nanoparticles with antibacterial activity, the most accepted mechanism of action of how nanoparticles inhibit bacterial proliferation is by their union to the negatively charged cell surface, thus satisfying the denaturation of proteins and causing cell death [36]. AgNPs are recognized for their antibacterial activity [37]. In another work and using SAXS as a characterization technique of size distribution, Wuithschick et al. developed the first selective post-synthesis method of fractionation by the size of AgNPs synthesized in the same batch [35]. The evaluation of the antibacterial activity of the fractions obtained showed that a smaller size of the AgNPs enhances the antibacterial activity.

3.3.2 Nanoparticles and Cancer Therapies

Another biomedical application of nanoparticles has been their use in the detection, diagnosis, and targeted drug delivery of cancer cells [38, 39]. The use of gold nanoparticles (AuNPs) in this field has expanded due mainly to its relatively easy surface functionalization and its high specificity and permeability toward tumor cells [40]. The high specificity and preferential accumulation in tumor cells present in AuNPs and other metallic nanoparticles are strongly dependent on their size [41]. The above is due to the different permeabilization presented by healthy and tumoral tissues [40]. By passive targeting, nanoparticles can enter into tumor cells due to the enhanced permeability and retention (EPR) phenomenon that they present [42]. Le Goas et al. synthesized and characterized AuNPs functionalized with various polymers and copolymers to study their potential use as radiosensitizers in cancer therapy [43]. The AuNPs and functionalized AuNPs were irradiated with multiple doses of gamma rays to evaluate their capacity to generate reactive species against tumor cells. Complementing measures of SAXS and SANS, they were able to fully characterize the effect of gamma irradiation on the structures of these nanoparticles [43]. SAXS measurements showed that the metal core of the AuNPs did not undergo structural modifications and SANS measurements showed that only the polymer crown of these AuNPs was affected by radiation doses. Other nanoparticles used as cancer treatment and characterized by SAXS and SANS have been the copper nanoparticles (CuNPs) [44, 45] and AgNPs [46].

3.3.3   Proenzyme-like Nanoparticle

A proenzyme is an enzyme temporary inactivated physically or chemically, which upon adequate external stimulus recovers its catalytic activity [47]. One of the few examples of materials with this property for a future biomedical application in biological systems with high oxidative stress is the nanoparticles of $Ce(OH)_3$. Bohn et al. synthesized these nanoparticles and evaluated their catalytic activity against the dismutation reaction of superoxide ($O_2{}^-$) to molecular oxygen (O_2) and hydrogen peroxide (H_2O_2) [48]. These nanoparticles acquire superoxide dismutase activity in the presence of H_2O_2. Upon contact with H_2O_2, the content of Ce(III) decreases, the nanoparticles of $Ce(OH)_3$ becoming nanoparticles of the type $CeO_{(2-x)}$ type, which present superoxide dismutase activity. To obtain this type of nanoparticles and that these were stable at physiological pH, an innovative strategy of synthesis was used, which consisted of using phytantriol, amyotrophic liquid crystal [48]. The characterization by SAXS allowed determining the phase structure of the systems. The SAXS curves displayed diffraction peaks, which confirmed that the samples were assembled as a lyotropic liquid crystalline structure. The SAXS results suggested that the nanoparticles were in the aqueous domains of the liquid crystalline systems.

3.4   X-ray Photoelectron Spectroscopy (XPS) and X-ray Absorption Spectroscopy (XAS)

X-ray photoelectron spectroscopy (XPS) is one of the main techniques used in the characterization of the biomaterial surface because of its responsibility for the success or failure of a biomaterial device [49]. In addition to providing a composition, it is possible to verify the chemical environment of the atoms of the sample surface [50]. XPS technique is considered a relatively non-destructive, quite mature method, increasingly used in different areas of biology, including cell, bacteria, and tissue analysis, as well as bioengineering [51]. This susceptible surface technique is frequently being used in the field of biomolecule characterization for application with proteins, peptides, lipids, mucins, enzymes, and DNA [52, 53].

XPS is a technique based on the photoelectric effect, in which the surface to be analyzed is irradiated with soft X-ray photons (conventional sources of Mg Kα

(1253.6 eV) and Al Kα (1486.6 eV), or synchrotron radiation is used whose energy can be chosen within specific intervals) [54]. When a photon of energy $h\nu$ interacts with an electron at a level with binding energy (BE), the energy of the photon is entirely transferred to the electron, with the result of the emission of a photoelectron with KE (kinetic energy). Figure 3.2 illustrates the photoelectric effect for an isolated atom, where a photoelectron of the electronic layer K is emitted [55]. In this context, when the Fermi levels of the sample and the analyzer are leveled, the sample work function (φ) is replaced by the analyzer's work function (φ). An additional work (φ) is required to altogether remove the electron from the material to the so-called vacuum level. In addition, (3.10) indicates that any change in the BEs will be reflected in the KEs, which means that changes in the chemical environment of an atom can be studied with changes in photoelectronic energies, supplying chemical information. XPS can analyze all the elements of the periodic table except for hydrogen and helium.

The kinetic energy of the photoelectron emitted can be written as:

$$KE = h\nu - BE - \varphi \tag{3.10}$$

where KE kinetic energy of the photoelectron; $h\nu$: incident photon energy; BE: binding energy; φ: sample work function.

For the analysis of the photoelectrons, hemispherical analyzers are used, which are made up of two metal half-spheres, among which there is a potential difference. The sample should be kept in an ultrahigh vacuum chamber to keep the sample surface unchanged during data collection and to minimize the inelastic scattering [56]. An electric field is generated between the half-beads of the analyzer due to the potential difference applied. This electric field is responsible for selecting the electrons that arrive at the analyzer with specific kinetic energy. At the input of the analyzer, there are electrostatic lenses that have the function of focusing and delaying the photoelectrons up to selected energy, called passing energy because they have kinetic energy values too high to be deflected in the path inside the analyzer. At the end of the electron trajectory, there is a multichannel detector, where each channel is

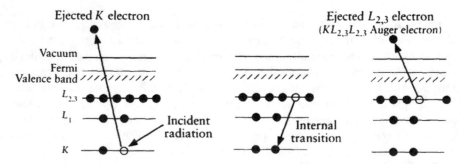

Fig. 3.2 Representation of the photoelectron emission process and Auger effect. Reproduced with permission from [55]

responsible for a range of energy. Thus, each electron detected in a specific energy interval adds a count in the channel. Hence, the photoemission spectrum is obtained as a function of kinetic energy (Fig. 3.2).

The primary information obtained in XPS measurements is the composition and the chemical state of the atoms on the surface of the samples [56]. The intensity of a given peak in the photoemission spectrum is proportional to the concentration of corresponding atoms in the sample [57]. The XPS spectrum is formed by a series of peaks, on a background, produced by the inelastic collisions of the electrons that lose most of their energy before leaving the sample. The energy of the X-rays used limits the electronic levels of each element that can undergo photoionization, so they only photoemit the levels with the highest useful absorption section of the incident photons. For each chemical element, a set of peaks is always observed at characteristic energies that allow it to be identified, usually using the most intense photoelectronic peak or central peak, to perform the quantitative and chemical state analysis. In addition to this type of peaks, others may appear due to other processes (Auger peaks), making the XPS spectrum sometimes complicated (Fig. 3.3).

The XPS chemical shift can be described as (3.11):

$$E_b^i = k_{qi} + \sum_j \frac{q_j}{r_{ij}} + E_b^{\text{ref}} \tag{3.11}$$

where E_b^i determine the XPS binding energy, k proportionality constant, q_i charge of the atom, $\sum_j q_j / r_{ij}$ sum of Coulomb contributions from the neighbors at distance r_{ij}, E_b^{ref} reference energy (e.g., metal). The inelastic mean free path corresponds to the average distance that an electron with specific energy can traverse in a sample before it undergoes an inelastic collision [56]. Figure 3.4 shows the dependence of the inelastic mean free path with the kinetic energy of the electrons; this curve is also

Fig. 3.3 XPS spectrum of a standard copper powder sample acquired by Al Kα and Mg Kα sources to distinguish between XPS and Auger peaks. Reproduced with permission from [55]

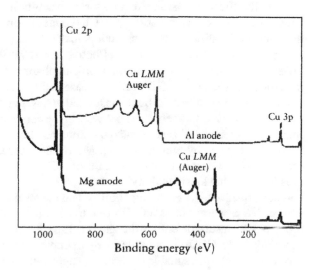

Fig. 3.4 Mean inelastic free path (λ) as a function of the kinetic energy of the electrons for different materials. The continuous curve represents the theoretical prediction, while the points are experimental data for some elements. Reproduced with permission from [58]

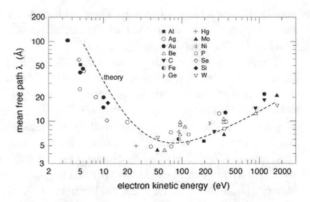

known as the universal mean free path curve [55, 56]. The XPS technique is widely used in the characterization of nanometric materials because it has a high sensitivity to the surface of the sample due to the short inelastic free path of the electrons inside the solid.

3.5 XPS for Biomaterial Characterization with Bio-pharm-med Applications

3.5.1 Biomaterial Biocompatibility and Their Use as Drug Delivery

The scientific and technological development in the field of biomedicine seeks to increase the effectiveness of medicines through new designs in their supply or administration systems of active substances. Challenges in drug delivery systems involve the effective localization of therapeutic substances in preselected locations, and in concentrations that maximize their effectiveness [59, 60]. Encapsulation of active materials is widely used today in the medical, pharmaceutical, and cosmetic industries for the development of controlled release delivery systems [59]. Many therapeutically active molecules must be encapsulated in a system that prevents their degradation and guarantees their effective delivery in the corresponding biological medium. Silk fibroin biomaterial-functionalized carbon nanotubes for high water dispersibility were presented by Sun et al. [61]. Multiwalled carbon nanotubes (MWCNTs) functionalized with water-soluble silk fibroin (SF) were prepared via chemical modification, showing low in toxicity. Additionally, a methyl thiazolyl tetrazolium assay was performed to assess biocompatibility, and the results indicated the desirable biocompatible properties of MWCNTs due to SF functionalization (amino-terminated groups). XPS results determine the chemical species introduced by the modification, especially C=O, COOH groups, that help to improve their water dispersibility, transforming MWCNTs in suitable materials for applications in biological and biomedical systems and devices. In this field, Pawlik et al. reported the use of amorphous TiO_2 (am) layers modified with 3-aminopropyltriethoxysilane (APTES), as drug delivery systems and scaffolds for cell culture [62]. However, the use of NaOH enhanced the ibuprofen release from TiO_2 am layers, improving the metabolic activity of adhered cells compared with the non-modified and NaOH-modified TiO_2 layers (Nam). XPS showed a substantial difference before (TiO_2-am, and Nam) against APTES modification (ANam). The presence of Si 2p and N 1s peaks and a slight increase in the intensity of the C1s peak for the sample modified with APTES, confirming that the two-step modification (NaOH and APTES) procedure was applied effectively [62].

Urothelial diseases (UD) were studied using bilayer swellable drug-eluting ureteric stent (BSDEUS) [63]. This new drug delivery platform technology was applied for the treatment of UD, such as strictures and carcinomas. BSDEUS consists of a stent spray-coated with a polymeric drug containing polylactic acid-co-caprolactone (PLC) layer which is overlaid by a swellable polyethylene glycol diacrylate (PEGDA)-based hydrogel. XPS measurements were performed on the drug-and-PLC-coated stents to evaluate the effect of plasma treatment times on the surface oxidation, indicating an increase in oxygen content at longer plasma duration.

3.5.2 Biomaterials Applied as Antimicrobials and Cancer Agents

Different studies have evaluated the antimicrobial effect of various materials of biological origin; among these, chitosan, a natural polymer, biodegradable, non-toxic in moderate concentrations, and bearer of some antimicrobial activity stands out as one of the most recurrently used materials [64]. The antimicrobial effect of chitosan is attributed to the chelating capacity and the presence of a positively charged amino group that can interact with the compounds of opposite charge and which are present on the surface of the microorganisms [60, 64].

Adjnik et al. examined the physicochemical properties of functionalized silicone materials with antimicrobial coatings and chitosan over the surface [65]. The authors found a strong influence of the use of silicones by drug-embedded chitosan nanoparticles. Furthermore, in vitro drug release testing was used to follow the desorption kinetics and antimicrobial properties were tested by a bacterial cell count reduction assay using the standard gram-positive bacteria *Staphylococcus aureus*. XPS analysis shows that the presence of nitrogen indicates that the chitosan nanoparticles with the incorporated drug were attached to the silicone material, which confirms that they are suitable materials to being studied as an antimicrobial agent. Antimicrobial peptides (AMPs) were analyzed and presented by He et al, using a novel antimicrobial surface by AMPs with improved stability [66]. The authors reported that antimicrobial peptide HHC36 with L-propargylglycine (PraAMP) to improve its salt-tolerant activity and integrate this PraAMP onto the spacer molecule by click chemistry. XPS confirmed the success of the immobilization, observing that structural changes occurring during the two synthetic processes, which improved the stability, particularly an enhanced enzymolysis tolerance.

Duta et al. studied and demonstrated that the post-deposition thermal treatment of new bioactives, antimicrobial and adherent coatings of nanostructured carbon double-reinforced with silver promotes a higher amorphization and an increase of the sp^2/sp^3 C species ratio [67]. Moreover, no contaminants were detected inside deposited structures. XPS was also applied to determine the structural composition of combinatorial maps fabricated from chitosan and biomimetic apatite powders for orthopedic applications, validating the chemical composition of the C-MAPLE thin films [67]. Accordingly, the survey spectra acquired on the surface of CHT-BmAp samples exhibit: C 1s, O 1s, N 1s, Ca 2s, Ca 2p, P 2s, and P 2p photoelectron peaks. The atomic ratio (N 1s/N 1s + Ca 2p) is decreasing from CHT to BmAp compounds, indicating the composition gradient of combinatorial layers. On the other hand, the Ca 2p/P 2p ratio for CHT-BmAp compounds varied between 0.9 and 1.03 along the sample length [67].

Recently, Fojtu et al. report the synthesis of black phosphorus nanoparticles (BPNPs) and exploration of their applicability in targeted drug delivery [68]. BPNPs are loaded with platinum agents—cisplatin and oxaliplatin and explored the applicability of BP loaded with two commercial platinum anticancer agents, cisplatin (CP) and oxaliplatin (OP). The binding abilities of cisplatin and oxaliplatin to the FLBP surface were examined by wide scan XPS. The peaks from 72.5 to 76.5 eV confirmed the presence of Pt–P bonds in CP- and OP-bound BP. The binding between OP and BP was much stronger than that between CP and BP because of the presence of 1,2-diaminocyclohexane and a new oxalate group in OP. Thus, BP–OP presented higher cytotoxicity when assessed in the human ovarian cancer cell line A2780 after treatment for 24 h. Todea et al. investigated the used XPS to analyze new solid forms of antineoplastic agent 5-fluorouracil (5-FA) with anthelmintic piperazine [69]. The authors emphasized that XPS allows to elucidated information on the atomic environments of 5-FA, where the deconvolution of N 1s core level allows to determine undoubtedly that the protonated (salt) from hydrogen-bonded (co-crystal) nitrogen species [69].

3.6 X-ray Absorption Spectroscopy (XAS)

The X-ray absorption spectroscopy technique allows studying the local atomic structure around an individual atom [70]. This technique has been used in the characterization of nanometric materials to provide electronic information (oxidation state and density of unoccupied states) and structural (interatomic distances, coordination number, and structural disorder) around a specific atom [71]. An essential point in the use of the XAS technique is the possibility of in situ measurements, that is, measurements performed during reactions at different temperatures and with varying atmospheres of gas interacting with the sample.

The use of the XAS technique requires the use of high-brightness, monochromatic radiation with enough energy for each element that is analyzed (chosen by varying the angle of incidence of the polychromatic radiation in a monochromator crystal) [71]. Hence, the synchrotron radiation emitted when charged relativistic particles (usual electrons with energy between 100 MeV and 10 GeV) are deflected by magnetic fields is essential for performing XAS measurements. This radiation covers a wide range of the electromagnetic spectrum, from infrared to hard X-rays. Synchrotron radiation has a brightness thousands of times higher than the radiation produced by conventional X-ray tubes. The XAS technique consists of measuring the absorption coefficient of X-rays as a function of the energy of the incident monochromatic beam. The X-ray absorption coefficient measured in the XAS experiment can be described by the Beer–Lambert Law (3.12):

$$I = I_0 e - \mu x \qquad (3.12)$$

where I—transmitted beam intensity; I_0—incident beam intensity; μ—material absorption coefficient; x—sample thickness. When a beam of X-rays of intensity I_0 passes through a material of thickness x, the transmitted beam has its intensity reduced due to various phenomena of X-ray interaction with matter. However, for this energy range of incident radiation, the most important aspect is the absorption of photons through the photoelectric effect. The absorption coefficient presents direct dependence on the energy of the incident X-rays, decreasing with the increase in the energy of the incident photons. It can be observed that in specific energies there is a rapid increase in the coefficient of absorption. The region in energy where this increase occurs is called the absorption edge and is characteristic of each element.

Considering an isolated absorber atom, that is, without any near neighbor atom, the theoretical absorption spectrum after the edge must be a smooth-type drop without oscillations [72]. However, when the absorber atom has several close neighbors (Fig. 3.5), oscillations are observed after the absorption edge. These oscillations occur due to the scattering of the photoelectrons by the atoms neighboring the absorber atom, creating interferences between the backscattered wave function by the neighboring atoms with the wave function emitted by the absorber. The interferences may be constructive or destructive and produce the oscillations observed after the absorption edge.

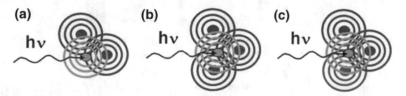

Fig. 3.5 Schematic of the EXAFS effect: The photoelectron ejected from the X-ray absorbing atom is scattered by different configurations of neighbors. Adapted with permission from [72]

The absorption of a photon occurs when the energy of the incident beam ($h\upsilon$) is greater than or equal to the bonding energy of the electrons (E_b), that is, $h\upsilon \geq E_b$. When the incident beam has energy $h\upsilon$ smaller than the bonding energy (E_b) of an electron, or, i.e., $h\upsilon < E_b$, the emission of photoelectrons of this layer does not occur [72, 73]. However, when the energy of the incident beam is higher than the Eb of the electrons ($h\upsilon > E_b$), the photoelectron has kinetic energy (E_k) and is scattered by the potential of the neighboring atoms. The photoelectron emitted can be considered as a spherical wave of wavelength (3.13):

$$\lambda = 2\pi k \tag{3.13}$$

where k is the photoelectron wave vector (3.14)

$$k = \frac{2\pi}{h}\sqrt{2m_e E_k} = \frac{2\pi}{h}\sqrt{2m_e(h\upsilon - E_b)} \tag{3.14}$$

$$\chi(k) = \sum_j Aj(k)\sin\left(2kr_j + \Phi j(k)\right) \tag{3.15}$$

m is the mass of the electron and $\hbar = h/2\pi$ (h is the Planck constant) [72]. The absorption spectrum is divided into two distinct energy regions, as shown in Fig. 3.6 as follows:

X-ray absorption near edge structure (XANES) involves a range of energy near the absorption edge and up to 50 eV after the edge [72, 74]. In the XANES region, the wavelength of the photoelectron is of the order of the interatomic distances, so the free inelastic mean free path is broad, and multiple scattering occurs before it

Fig. 3.6 Transmission spectrum at the Co K-edge as a function of the photon energy; d is the thickness of the metal foil; $\mu(E)$ is the absorption coefficient. Reproduced with permission from [74]

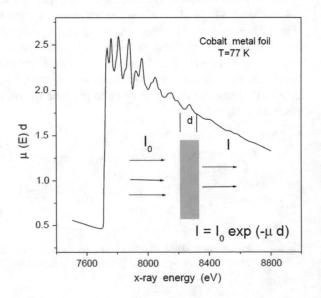

returns to the absorptive atom (Fig. 3.6). In addition to the multiple scattering, the XANES region allows transitions to unoccupied levels, so this technique will enable us to obtain information about the oxidation state and the density of empty states of the absorber atom.

The observed oscillations in the EXAFS region depend on the photoelectron wavelength, the number of neighbors, atomic number, and position of the neighboring atoms [75]. These oscillations of the EXAFS χ (k) region can be described as the sum of the contribution from the scattering of the photoelectron in different layers j:

$$A_j(k) = N_j \frac{e^{-2r_j/\lambda(k)}}{kr_j^2} S_0^2(k) F_j(k) e^{-2k^2\sigma_j^2} \tag{3.16}$$

where, N_j—number of atoms present in layer j (coordination numbers); S_0^2—amplitude reduction factor; k—photoelectron wave vector; $F_j(k)$—backscatter amplitude of the photoelectron with wave vector k (chemical sensitivity); σj—Debye–Waller factor; r_j—distance between the scattering atom j and the absorber atom; Φ_j k—phase shift due to scattering, and $\lambda(k)$ inelastic mean free path of the photoelectron [75].

The term $e^{-2r_j/\lambda(k)}$ has a dependence on the inelastic mean free path and represents the attenuation of the wave associated with the photoelectron when it travels through a solid. The exponential $e^{-2k^2\sigma_j^2}$ provides the sample disorder through the Debye–Waller factor, σ_j, which includes contributions of thermal and structural disorder. Factor 2 in the exponentials and in the sine indicates that the photoelectron traverses a closed path between the scattering atom and the absorber. The term $1/r^2$ reflects the fact that the photoelectron ejected from the atom behaves like a spherical wave, where the intensity decreases with the square of the distance. The argument of the sine function depends on k, R, and $\Phi_j k$; the value of $\Phi_j k$ is simulated. The energy loss processes are considered by the amplitude reduction factor S_0^2. The scattering amplitude $F_j(k)$ is the probability of the wave function of the photoelectron with wave vector k being scattered at a given angle by the neighboring atoms. Thus, the analysis of EXAFS oscillations allows the obtaining of structural parameters such as the number of neighboring atoms, Debye–Waller factor, and interatomic distances [72, 75].

3.6.1 XAS for Biomaterial Characterization with Bio-pharm-med Applications

A vast number of reports have been used X-ray absorption spectroscopy to characterize and obtain information about the chemical state and neighborhood of the NPs with strong bio-pharm-med applications [76, 77]. Recently, Rubina et al. reported the use of AgNPs to understand the multifunctional interaction of the system collagen—chitosan material containing silver nanoparticles and non-steroid anti-inflammatory

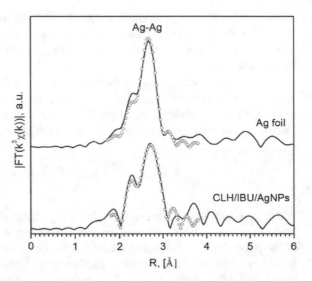

Fig. 3.7 Fourier transforms of EXAFS spectra for CLH/IBU/AgNPs: experimental (line) and best-fit theoretical (open circles) curves. Reproduced with permission from [78]

drug ibuprofen [78]. In this work, they proposed the in vitro release of ibuprofen in phosphate buffer, where the drug release to the solution is governed by Fickian diffusion. The use of AgNPs does not affect the diffusion mechanism, showing that these materials have no specific interactions of AgNPs with a collagen-chitosan scaffold and ibuprofen detected (Fig. 3.7).

Ma et al. examined the atomically dispersed Fe–N_4 sites anchored on N-doped porous carbon materials (Fe-SAs/NC) [78]. These materials can mimic two antioxidative enzymes of catalase and superoxide dismutase and therefore serve as a bifunctional single-atom-based enzyme for scavenging reactive oxygen species (ROS) to remove excess ROS generated during oxidative stress in cells. Using XANES and EXAFS, it was possible to investigate the local structure of Fe-SAs/NC at the atomic level, suggesting that the valence state of Fe in Fe-SAs/NC was between +2 and +3. Also, the obtained coordination number of Fe in the Fe-SAs/CN was about 4, and the bond length between Fe atoms and surrounding coordination atoms (N) was 2.01 Å, corresponding to the first coordination shell of Fe–N.

XANES was also used to describe a multifunctional magnetic drug delivery system made up of iron oxide nanoparticles (IONPs) and a Pluronic F127 shell, to carry doxorubicin (DOX) for neuroblastoma treatment. Mdlovu et al. analyzed to characterize the components of this system, where XANES was performed to understand the fine structure arrangement, electronic configuration, stereochemistry, and oxidation states in terms of bond distance and coordination number of Fe, demonstrated an absorbance feature (Fe = 7112 eV) of a 1s to 3d transition. The 3-(4,5-dimethylthiazol-2-yl)-2,5-diphenyltetrazolium bromide (MTT) results proved that the prepared nanocarriers were non-toxic when tested on *BE*–2–M17 cells [79].

During the last decade, several studies have reported the development of calcium phosphate varieties, scaffolds with random or ordered porosity for anticancer and antibacterial treatment. Within this family, a new generation of biomaterials with multifunctional europium(III)-doped Hap scaffolds has shown remarkable developments [79]. The luminescent multifunctional biomaterials show potential for use in various biomedical applications such as smart drug delivery, bioimaging, and photothermal therapy [80]. XANES and EXAFS showed that Eu^{+3} ions were incorporated preferentially at the Ca site with local charge compensating via oxygen interstitial ion for samples calcined under air at 450 °C/4 h °C. This effect was due to two possible phenomena: (i) reduction in the average coordination number of Eu(III) ions on the samples produced in the presence of CTAB, or (ii) increase in the Debye–Waller factor that can be associated with an increase in local intrinsic disorder induced by CTAB interaction with cationic Ca^{2+} or PO_4^{3-} ions [81].

3.7   Conclusions and Outlook

In this chapter, a short outline of the theory of SAXS, SANS, XPS, XANES, and EXAFS, with some examples of state-of-the-art experiments and applications were presented. In brief, neutron scattering determines the structure with unique possibilities to understand the atomic distribution at microscopic level and adsorbed molecules in different biomedical applications. XPS offers the possibility to study the surface sense with elemental composition and distribution of oxidation states, with an analytical potential for synthesis and characterization of new biomaterials. Synchrotron-based speciation with XAS offers unique features for the analysis of chemical element species in biological samples. The use of synchrotron radiation measures (XANES, EXAFS) highlighted the importance of the advanced techniques to characterize these new biomaterials, especially under reaction conditions. It is possible to explore the properties of atoms located at the surface or isolated with the aim to understand the behavior under biological conditions.

Acknowledgements The authors are grateful to FONDECYT Iniciación N°11170879. RZB thanks Dicyt-USACH for a postdoctoral grant N021841AL_POSTDOC.

Disclosure All authors have read and approved the final version.

References

1. McNamara K, Tofail SAM. Nanosystems: the use of nanoalloys, metallic, bimetallic, and magnetic nanoparticles in biomedical applications. Phys Chem Chem Phys. 2015;17(42):27981–95.
2. Mout R, Moyano DF, Rana S, Rotello VM. Surface functionalization of nanoparticles for nanomedicine. Chem Soc Rev. 2012;41(7):2539–44.
3. Huang X, Li L, Liu T, Hao N, Liu H, Chen D, Tang F. The shape effect of mesoporous silica nanoparticles on biodistribution, clearance, and biocompatibility in vivo. ACS Nano. 2011;5(7):5390–9.

4. Spicer CD, Jumeaux C, Gupta B, Stevens MM. Peptide and protein nanoparticle conjugates: versatile platforms for biomedical applications. Chem Soc Rev. 2018;47(10):3574–620.
5. Rahman IA, Padavettan V. Synthesis of Silica nanoparticles by Sol-Gel: size-dependent properties, surface modification, and applications in silica-polymer nanocompositesa review. J Nanomater. 2012;2012.
6. Carlson C, Hussein SM, Schrand AM, Braydich-Stolle LK, Hess KL, Jones RL, Schlager JJ. Unique cellular interaction of silver nanoparticles: size-dependent generation of reactive oxygen species. J Phys Chem B. 2008;112(43):13608–19.
7. Woźniak A, Malankowska A, Nowaczyk G, Grześkowiak BF, Tuśnio K, Słomski R, Zaleska-Medynska A, Jurga S. Size and shape-dependent cytotoxicity profile of gold nanoparticles for biomedical applications. J Mater Sci. 2017;28(6):1–11.
8. Alford A, Kozlovskaya V, Kharlampieva E. Small angle scattering for pharmaceutical applications: from drugs to drug delivery systems. In: Chaudhuri B, Muñoz IG, Qian S, Urban VS, editors. Biological small angle scattering: techniques, strategies and tips. Singapore: Springer; 2017. p. 239–62.
9. Li T, Senesi AJ, Lee B. Small angle X-ray scattering for nanoparticle research. Chem Rev. 2016;116(18):11128–80.
10. Di Cola E, Grillo I, Ristori S. Small angle X-ray and neutron scattering: powerful tools for studying the structure of drug-loaded liposomes. Pharmaceutics. 2016;8(2):1–16.
11. Nawroth T, Johnson R, Krebs L, Khoshakhlagh P, Langguth P, Hellmann N, Goerigk G, Boesecke P, Bravin A, Duc GL and others. Target nanoparticles for therapy-SANS and DLS of drug carrier liposomes and polymer nanoparticles. J Phys Conf Ser. 2016;746(1):28–31.
12. Chu B, Liu T. Characterization of nanoparticles by scattering techniques. J Nanopart Res. 2000;2(1):29–41.
13. Becker J. Plasmons as sensors. Berlin, Heidelberg: Springer; 2012.
14. Roe PMSRJ, Roe RJ. Methods of X-ray and neutron scattering in polymer science. New York: Oxford University Press; 2000. p. 331.
15. Kempkens H, Uhlenbusch J. Scattering diagnostics of low-temperature plasmas (Rayleigh scattering, Thomson scattering, CARS). Plasma Sour Sci Technol. 2000;9(4):492–506.
16. In the case of neutrons, these are dispersed by the nucleus.
17. Craievich AF. Small-angle X-ray scattering by nanostructured materials. In: Klein L, Aparicio M, Jitianu A, editors. Handbook of sol-gel science and technology: processing, characterization and applications. Cham: Springer International Publishing; 2018. p. 1185–230.
18. Dmitri IS, Michel HJK. Small-angle scattering studies of biological macromolecules in solution. Rep Prog Phys. 2003;66(10):1735.
19. Jiao Y, Akcora P. Understanding the role of grafted polystyrene chain conformation in assembly of magnetic nanoparticles. Phys Rev E. 2014;90(4):1–9.
20. Bonini M, Fratini E, Baglioni P. SAXS study of chain-like structures formed by magnetic nanoparticles. Mat Sci Eng C. 2007;27(5–8 SPEC. ISS.):1377–81.
21. Boldon L, Laliberte F, Liu L. Review of the fundamental theories behind small angle X-ray scattering, molecular dynamics simulations, and relevant integrated application. Nano Rev. 2015;6(1):25661.
22. Londoño OM, Tancredi P, Rivas P, Muraca D, Socolovsky LM, Knobel M. Small-angle X-ray scattering to analyze the morphological properties of nanoparticulated systems. Cham: Springer International Publishing; 2018. p. 37–75.
23. Agbabiaka A, Wiltfong M, Park C. Small angle X-ray scattering technique for the particle size distribution of nonporous nanoparticles. J Nanoparticles. 2013;2013:1–11.
24. Bender P, Bogart LK, Posth O, Szczerba W, Rogers SE, Castro A, Nilsson L, Zeng LJ, Sugunan A, Sommertune J and others. Structural and magnetic properties of multi-core nanoparticles analysed using a generalised numerical inversion method. Sci. Reports. 2017;7:1–14.
25. Ristori S, Grillo I, Lusa S, Thamm J, Valentino G, Campani V, Caraglia M, Steiniger F, Luciani P, De Rosa G. Structural characterization of self-assembling hybrid nanoparticles for Bisphosphonate delivery in tumors. Mol Pharm. 2018;15(3):1258–65.

26. He W, Yan J, Sui F, Wang S, Su X, Qu Y, Yang Q, Guo H, Ji M, Lu W and others. Turning a Luffa protein into a self-assembled biodegradable nanoplatform for multitargeted cancer therapy. ACS Nano 2018;12(11):11664–77.
27. Lakshminarayanan R, Ye E, Young DJ, Li Z, Loh XJ. Recent advances in the development of antimicrobial nanoparticles for combating resistant pathogens. Adv Healthc Mat. 2018;7(13):1–13.
28. García I, Henriksen-Lacey M, Calvo J, De Aberasturi DJ, Paz MM, Liz-Marzán LM. Size-dependent transport and cytotoxicity of mitomycin-gold nanoparticle conjugates in 2D and 3D Mammalian cell models. Bioconjugate Chem. 2019;30(1):242–52.
29. de Souza ME, Verdi CM, de Andrade ENC, Santos RCV. Chapter 12—antiviral and antimicrobial (antibacterial) potentiality of nano drugs. In: Mohapatra SS, Ranjan S, Dasgupta N, Mishra RK, Thomas S, editors. Applications of targeted nano drugs and delivery systems. Elsevier; 2019. pp. 327–42.
30. Spagnol C, Fragal EH, Pereira AGB, Nakamura CV, Muniz EC, Follmann HDM, Silva R, Rubira AF. Cellulose nanowhiskers decorated with silver nanoparticles as an additive to antibacterial polymers membranes fabricated by electrospinning. J Coll Interf. Sci. 2018;531:705–15.
31. Cagno V, Andreozzi P, D'Alicarnasso M, Silva PJ, Mueller M, Galloux M, Goffic RL, Jones ST, Vallino M, Hodek J and others. Broad-spectrum non-toxic antiviral nanoparticles with a virucidal inhibition mechanism. Nat Mater. 2018;17(2):195–203.
32. De Souza E, Silva JM, Hanchuk TDM, Santos MI, Kobarg J, Bajgelman MC, Cardoso MB. Viral inhibition mechanism mediated by surface-modified silica nanoparticles. ACS Appl Mater Interf. 2016;8(26):16564–72.
33. Sokolowski M, Bartsch C, Spiering VJ, Prévost S, Appavou MS, Schweins R, Gradzielski M. Preparation of polymer brush grafted anionic or cationic silica nanoparticles: systematic variation of the polymer shell. Macromolecules. 2018;51(17):6936–48.
34. Yi Z, Dumée LF, Garvey CJ, Feng C, She F, Rookes JE, Mudie S, Cahill DM, Kong L. A new insight into growth mechanism and kinetics of mesoporous silica nanoparticles by in situ small angle X-ray scattering. Langmuir. 2015;31(30):8478–87.
35. Wuithschick M, Paul B, Bienert R, Sarfraz A, Vainio U, Sztucki M, Kraehnert R, Strasser P, Rademann K, Emmerling F and others. Size-controlled synthesis of colloidal silver nanoparticles based on mechanistic understanding. Chem Mater. 2013;25(23):4679–89.
36. Varier KM, Gudeppu M, Chinnasamy A, Thangarajan S, Balasubramanian J, Li Y, Gajendran B. Nanoparticles: antimicrobial applications and its prospects. In: Naushad M, Rajendran S, Gracia F, editors. Advanced nanostructured materials for environmental remediation. Cham: Springer International Publishing; 2019. p. 321–55.
37. Lara HH, Ayala-Nuñez NV, Ixtepan-Turrent L, Rodriguez-Padilla C. Mode of antiviral action of silver nanoparticles against HIV-1. J Nanobiotechnol. 2010;8:1–10.
38. Chang ZM, Wang Z, Shao D, Yue J, Xing H, Li L, Ge M, Li M, Yan H, Hu H and others. Shape engineering boosts magnetic mesoporous silica nanoparticle-based isolation and detection of circulating tumor cells. ACS Appl Mater Interf 2018;10(13):10656–63.
39. Singh P, Pandit S, Mokkapati VRSS, Garg A, Ravikumar V, Mijakovic I. Gold nanoparticles in diagnostics and therapeutics for human cancer. Int J Mol Sci 2018;19(7).
40. Sharma A, Goyal AK, Rath G. Recent advances in metal nanoparticles in cancer therapy. J Drug Targ. 2018;26(8):617–32.
41. Dhandapani R, Sethuraman S, Subramanian A. Nanohybrids—cancer theranostics for tiny tumor clusters. J Control Release. 2019;299:21–30.
42. Muhamad N, Plengsuriyakarn T, Na-Bangchang K. Application of active targeting nanoparticle delivery system for chemotherapeutic drugs and traditional/herbal medicines in cancer therapy: a systematic review. Int J Nanomed. 2018;13:3921–35.
43. Le Goas M, Paquirissamy A, Gargouri D, Fadda G, Testard F, Aymes-Chodur C, Jubeli E, Pourcher T, Cambien B, Palacin S and others. Irradiation effects on polymer-grafted gold nanoparticles for cancer therapy. ACS Appl Biomater. 2019;2(1):144–54.
44. Din MI, Arshad F, Hussain Z, Mukhtar M. Green adeptness in the synthesis and stabilization of copper nanoparticles: catalytic, antibacterial, cytotoxicity, and antioxidant activities. Nanoscale Res Lett. 2017;12.

45. Spencer E, Kolesnikov A, Woodfield B, Ross N. New insights about CuO nanoparticles from inelastic neutron scattering. Nanomaterials. 2019;9(3):312.
46. Spinozzi F, Ceccone G, Moretti P, Campanella G, Ferrero C, Combet S, Ojea-Jimenez I, Ghigna P. Structural and thermodynamic properties of nanoparticle-protein complexes: a combined SAXS and SANS study. Langmuir. 2017;33(9):2248–56.
47. Esmaeilzadeh P, Köwitsch A, Liedmann A, Menzel M, Fuhrmann B, Schmidt G, Klehm J, Groth T. Stimuli-responsive multilayers based on thiolated polysaccharides that affect fibroblast cell adhesion. ACS Appl Mater Interf. 2018;10(10):8507–18.
48. Bohn DR, Lobato FO, Thill AS, Steffens L, Raabe M, Donida B, Vargas CR, Moura DJ, Bernardi F, Poletto F. Artificial cerium-based proenzymes confined in lyotropic liquid crystals: synthetic strategy and on-demand activation. J Mater Chem B. 2018;6(30):4920–8.
49. Xu LC, Siedlecki CA. Protein adsorption, platelet adhesion, and bacterial adhesion to polyethylene-glycol-textured polyurethane biomaterial surfaces. J Biomed Mater Res. 2017;105(3):668–78.
50. Christo SN, Bachhuka A, Diener KR, Mierczynska A, Hayball JD, Vasilev K. The role of surface nanotopography and Chemistry on primary neutrophil and macrophage cellular responses. Adv Healthcare Mater. 2016;5(8):956–65.
51. Wang PY, Bennetsen DT, Foss M, Ameringer T, Thissen H, Kingshott P. Modulation of human mesenchymal stem cell behavior on ordered tantalum nanotopographies fabricated using colloidal lithography and glancing angle deposition. ACS Appl Mater Interf. 2015;7(8):4979–89.
52. Batista P, Castro PM, Madureira AR, Sarmento B, Pintado M. Recent insights in the use of nanocarriers for the oral delivery of bioactive proteins and peptides. Peptides. 2018;101:112–23.
53. Li SK, Liu ZT, Li JY, Chen AY, Chai YQ, Yuan R, Zhuo Y. Enzyme-free target recycling and double-output amplification system for electrochemiluminescent assay of Mucin 1 with MoS2 nanoflowers as Co-reaction accelerator. ACS Appl Mat Interf. 2018;10(17):14483–90.
54. Sezen H, Suzer S. XPS for chemical- and charge-sensitive analyses. Thin Solid Films. 2013;534:1–11.
55. Watts JF, Wolstenholme J. Electron spectroscopy: some basic concepts. An introduction to surface analysis by XPS and AES. Chichester, UK: Wiley & Sons, Ltd.; 2003. pp. 1–15.
56. Hofmann S. Introduction and outline. Auger- and X-ray photoelectron spectroscopy in materials science: a user-oriented guide. Berlin, Heidelberg: Springer; 2013. pp. 1–10.
57. Aziz M, Ismail AF. X-ray photoelectron spectroscopy (XPS). Elsevier; 2017. pp. 81–93.
58. Seah MP, Dench WA. Quantitative electron spectroscopy of surfaces: a standard data base for electron inelastic mean free paths in solids. Surf Interf Anal. 1979;1(1):2–11.
59. Fenton OS, Olafson KN, Pillai PS, Mitchell MJ, Langer R. Advances in biomaterials for drug delivery. Adv Mater. 2018;30(29):1–29.
60. Pokhriyal S, Gakkhar N, Bhatia A. Biomedical applications and toxicological effects of nanomaterials: a general approach. J Mat Sci Surf Eng. 2018;6(3):811–6.
61. Sun Y, Fu Y, Luo J, Wang R, Dong Y. Silk fibroin biomaterial-functionalized carbon nanotubes for high water dispersibility and promising biomedical applications. Textile Res J. 2019;89(7):1144–52.
62. Pawlik A, Socha RP, Hubalek Kalbacova M, Sulka GD. Surface modification of nanoporous anodic titanium dioxide layers for drug delivery systems and enhanced SAOS-2 cell response. Coll Surf B. 2018;171:58–66.
63. Lim WS, Chen K, Chong TW, Xiong GM, Birch WR, Pan J, Lee BH, Er PS, Salvekar AV, Venkatraman SS and others. A bilayer swellable drug-eluting ureteric stent: Localized drug delivery to treat Urothelial diseases. Biomaterials 2018;165:25–38.
64. Mitić Ž, Stolić A, Stojanović S, Najman S, Ignjatović N, Nikolić G, Trajanović M. Instrumental methods and techniques for structural and physicochemical characterization of biomaterials and bone tissue: a review. Mater Sci Eng C. 2017;79:930–49.
65. Ajdnik U, Zemljič LF, Bračič M, Maver U, Plohl O, Rebol J. Functionalisation of silicone by drug-embedded chitosan nanoparticles for potential applications in otorhinolaryngology. Materials. 2019;16(6):1–20.

66. He J, Chen J, Hu G, Wang L, Zheng J, Zhan J, Zhu Y, Zhong C, Shi X, Liu S and others. Immobilization of an antimicrobial peptide on silicon surface with stable activity by click chemistry. J Mater Chem B 2017;6(1):68–74.
67. Duta L, Ristoscu C, Stan GE, Husanu MA, Besleaga C, Chifiriuc MC, Lazar V, Bleotu C, Miculescu F, Mihailescu N and others. New bio-active, antimicrobial and adherent coatings of nanostructured carbon double-reinforced with silver and silicon by Matrix-Assisted Pulsed Laser Evaporation for medical applications. Appl Surf Sci. 2018;441:871–83.
68. Chia X, Fojtu M, Masar M. Black Phosphorus nanoparticles potentiate the anticancer effect of Oxaliplatin in ovarian cancer cell line. Adv Funct Mater. 2017;1701955(36):1–7.
69. Todea M, Simon S, Simon V, Eniu D. XPS investigation of new solid forms of 5-fluorouracil with piperazine. J Mol Struct. 2018.
70. Calvin S. XAFS for everyone. Boca Raton: Taylor & Francis Group; 2013. p. 459.
71. Charlet L, Manceau A. Chapter 4: Structure, formation and reactivity of hydrous oxide particles; insights from X-ray absorption spectroscopy. In: Buffle J, van Leeuwen HP, editors. Environmental particles 2. Boca Raton: CRC Press; 1993. p. 118–64.
72. Frenkel AI. Applications of extended X-ray absorption fine-structure spectroscopy to studies of bimetallic nanoparticle catalysts. Chem Soc Rev. 2012;41(24):8163–78.
73. Frenkel AI, Wang Q, Sanchez SI, Small MW, Nuzzo RG. Short range order in bimetallic nanoalloys: an extended X-ray absorption fine structure study. J Chem Phys. 2013;138(6):064202.
74. Faraci G. Cluster characterization by EXAFS spectroscopy. In: AIP 2002; 2002. pp. 173–177.
75. Koningsberger DC, Mojet BL, van Dorssen GE, Ramaker DE. XAFS spectroscopy; fundamental principles and data analysis. Topics Catal. 2000;10(3–4):143–55.
76. Azharuddin M, Zhu GH, Das D, Ozgur E, Uzun L, Turner APF, Patra HK. A repertoire of biomedical applications of noble metal nanoparticles. Chem Comm 2019:6964–96.
77. Kravtsova AN, Guda IV, Polozhentsev OE, Pankin IA, Soldatov AV. Xanes specroscopic diagnostics of the 3D local atomic structure of nanostructured materials. J Struct Chem. 2018;59(7):1691–706.
78. Rubina MS, Said-Galiev EE, Naumkin AV, Shulenina AV, Belyakova OA, Vasil'kov AY. Preparation and characterization of biomedical collagen–chitosan scaffolds with entrapped ibuprofen and silver nanoparticles. Pol Eng Sci. 2019:1–9.
79. Mdlovu NV, Mavuso FA, Lin KS, Chang TW, Chen Y, Wang SSS, Wu CM, Mdlovu NB, Lin YS. Iron oxide-pluronic F127 polymer nanocomposites as carriers for a doxorubicin drug delivery system. Coll Surf A. 2018;2019(562):361–9.
80. Su FX, Zhao X, Dai C, Li YJ, Yang CX, Yan XP. A multifunctional persistent luminescent nanoprobe for imaging guided dual-stimulus responsive and triple-synergistic therapy of drug resistant tumor cells. Chem Comm. 2019;55(36):5283–6.
81. Lima TARM, Valerio MEG. X-ray absorption fine structure spectroscopy and photoluminescence study of multifunctional europium (III)-doped hydroxyapatite in the presence of cationic surfactant medium. J Luminescence 2018;201(Iii):70–76.

Chapter 4
Computational Methodologies for Exploring Nano-engineered Materials

Ariela Vergara-Jaque, Matías Zúñiga and Horacio Poblete

Abstract Biomimetic nano-engineered materials have emerged as new potential additives for biomedical therapies. However, one of the most critical challenges that remain is the ability to produce responsive nanostructures that respond to external stimuli, enhance existing properties, and introduce new functionalities. In this regard, the use of computational methodologies to design, simulate, and visualize the interaction between biological substrates and nanostructures provides a powerful tool for better understanding structure/function. This chapter focuses on the use of molecular modeling and molecular dynamics (MD) methods to assist the design of bio-nanomaterials and characterize the structural aspects of the interaction between nanostructures and biological molecules. Computational simulations allow the analysis of the behavior of atoms and molecules for a period of time employing integrated mathematical and physical equations. Here, we describe how these theoretical methods are used to design and model nanomaterials in a rational way, as well as to evaluate its functionalization and association with drug-like compounds. Methodologies used in the field of computational nanotechnology include de novo modeling, parametrization, molecular dynamics simulations under functional conditions, binding free energy calculations, as well as future perspectives oriented to use reactive force field techniques.

A. Vergara-Jaque (✉) · M. Zúñiga · H. Poblete (✉)
Center for Bioinformatics and Molecular Simulation, Universidad de Talca,
2 Norte 685, 3460000 Talca, Chile
e-mail: arvergara@utalca.cl

H. Poblete
e-mail: hopoblete@utalca.cl

M. Zúñiga
e-mail: mzunigab@alumnos.utalca.cl

A. Vergara-Jaque · H. Poblete
Millennium Nucleus of Ion Channels-Associated Diseases (MiNICAD),
Universidad de Talca, Talca, Chile

4.1 Computer-Aided Design of Nanomaterials for Biomedical Applications

Advances in the design and controlled synthesis of nanomaterials have had a significant impact in the development of new therapeutics. Nanostructure-based technologies are used in modern medicine for drug and gene delivery [1], medical imaging [2, 3], cancer therapies [4], treatments for neurodegenerative diseases [5], tissue engineering [6], and regenerative medicine [7]. Identification of nanostructures with high biocompatibility and versatility that can be applied to the biomedical field, therefore, remains one of the main challenges of material engineering. In this regard, "computational nanotechnology" has become an important tool for understanding the design principles of systems at the nanoscale. The term "computational nanotechnology" was coined by Merkle in 1991 [8], emphasizing the benefits of using computer-aided design tools for the development of molecular manufacturing systems. An extensive variety of computational tools allows for behavior prediction and/or characterization of the properties of structures relevant to nanotechnology. Specialized software for modeling and simulation of nanomaterials are listed in Table 4.1. These varieties

Table 4.1 Computational tools for modeling and simulation of materials

Software	Purpose	Link
Materials studio	Simulation and modeling of materials	www.3dsbiovia.com
Atomistix ToolKit	Atomic-scale modeling and simulation of nanosystems	www.synopsys.com
Materials and Process Simulation (MAPS)	Building, simulation, and analysis of realistic models of all types of materials	www.scienomics.com
Material Exploration and Design Analysis (MedeA)	Atomistic simulation of materials	www.materialsdesign.com
MBN explorer and MBN studio	Computational modeling of molecular structure and dynamics of Meso-Bio-Nano (MBN) systems	www.mbnresearch.com
Software for Adaptive Modeling and Simulation Of Nanosystems (SAMSON)	Platform for computational nanoscience	www.samson-connect.net
SCIGRESS	Integrated platform for computational chemistry to support the development of materials	www.fqs.pl
Amsterdam Density Functional (ADF) suite	Computational chemistry platform for understanding and predicting structure and reactivity in chemistry and material science	www.scm.com

of software constitute powerful tools that can be used to design, simulate, and analyze nanostructures, either alone or interacting with external chemical or biological compounds.

Many issues surrounding the health risk of nanomaterials can be examined, prior to their synthesis and biological testing, using computational models [9, 10]. Structural and physiochemical properties such as size, shape, packing, chemical reactivity, and toxicity can be well described using computational algorithms [11 13]. In addition, *in silico* studies have shown tremendous potential for designing nanostructure-based therapies, reducing costs, and accelerating various steps in the development of biomaterials. Early efforts to design nanoparticles were focused on engineering materials with specific physical or chemical properties, but did not consider how the size, shape, and surface chemistry can affect the delivery and binding of nanostructures to target receptors. Polymer–drug conjugates and liposomes constitute the first generation of nanoparticles designed for nanoscale drug delivery purposes [14]. The first FDA-approved nanodrug, Doxil, was in fact formulated as a liposomal drug carrier to treat some types of cancer [15]. Molecular dynamics simulations and free energy calculations have been used to describe the specific interactions modulating the association of doxorubicin with different lipid membranes [16, 17]. Several other computational studies have rationalized the stability and effectivity of liposomal formulations [16, 18], highlighting the need to optimize the nanoparticle surfaces with functionalized groups in order to improve their stability and targeting in biological systems.

The second generation of nanomaterials is represented by functionalized structures with improved biocompatibility and pharmacokinetics. For this purpose, nanoparticle surfaces have been coated with a plethora of biomolecules, including lipids, peptides, carbohydrates, folic acid, DNA, antibodies, etc. However, polyethylene glycol (PEG) seems to be the most widely used strategy [19] (Fig. 4.1). Through coarse-grained molecular dynamics simulations, it was recently found

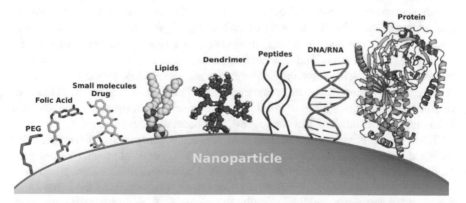

Fig. 4.1 Schematic representation of surface functionalization of nanoparticles. A wide variety of biomolecules are currently used to functionalize nanoparticles providing specific attributes for drug delivery, cancer therapy, imaging of cell and tissues, diagnostics, and sensing of target receptors

that a core–polyethylene glycol–lipid shell nanoparticle is able to carry hydrophilic molecules such as therapeutic compounds or imaging agents [20]. Molecular dynamics simulations have also allowed the characterization of PEG-conjugated gold nanoparticles designed for photothermal therapy [21, 22]. Metal nanoparticles exhibit beneficial properties for the treatment of malignant tumors by acting as radiosensitizers. In this regard, PEG functionalization is used to avoid clearance and increase half-life of these nanomaterials in the blood stream. PEG molecules can increase stability, reduce immunogenicity, and prolong circulation time of nanoparticles; however, its effectiveness depends on both the density of the PEG coating and length of the chains. As mentioned above, computational modeling has become an invaluable approach to model nanoparticles which allows the determination of many factors important in nanomaterial manufacturing, for instance, the optimal molecule size for coating nanomaterials and its concentration prior to the manufacturing process.

Nowadays, principles of nanoparticle design are focused on generating "intelligent" or "smart" nanomaterials able to response to external stimuli, be those physical, chemical, or biological. Particular differences between normal and damaged cells have led to the design of bio-nanomaterials stimulated by thermal gradients, pH variations, redox potentials, enzymatic activation, and magnetic fields among others [23]. These materials constitute the third generation of nanoparticles, which may be responsive to two or more combinations of different stimuli. Computational modeling and simulation methods, beyond allowing the study of structural and physiochemical properties of nanomaterials, provide tools to mimic physiological conditions and evaluate nanoparticles stability under variations of temperature, pressure, ion concentrations, pH, and electric fields. Atomistic models of the complexation of dexamethasone 21-phosphate with amine- and acetyl-terminated PAMAM dendrimer nanoparticles at different pH conditions were generated by MD simulations. These studies showed an increased affinity of the drug for acetyl-terminated PAMAM at low pH [24]. Temperature effects have been evaluated on carbon nanotube cap nucleation with different hydrocarbon species unraveling their growth mechanisms through combined reactive molecular dynamics and time-stamped force-bias Monte Carlo simulations [25]. Permeation of cationic gold nanoparticles across a phospholipid bilayer has been also been simulated under several intensities of external electric fields. Coarse-grained molecular dynamics simulation revealed that suitable ranges of electric fields allow the delivery of nanoparticles into the cell without disrupting the membrane after the permeation [26]. These examples denote the power of computational simulations to provide an atomic-level description of nanoparticle behavior under conditions that cannot easily be observed experimentally.

Table 4.2 summarizes the evolution of the three generations of nanoparticles designed for biomedical purposes. Conventional experimental techniques are generally used to design and explore the potential of these nanomaterials; however, simulation methods allow for guided experimentation and describe the behavior of bio-nanosystems more rapidly. Recently, the concept of "communicating nanoparticles" has also emerged [30, 31] (Fig. 4.2), which could be considered as a fourth generation of nanoparticles. Molecular modeling and MD simulation strategies, as

Table 4.2 Evolution of the design of nanomaterials, highlighting the properties of each of the three nanoparticle generations and the computational studies that have been carried out to characterize their functions.

Generation	Properties	Computational studies
First Drug / Liposome	– Passive nanostructures – Simple design – Vectors for drug solubilization – Biocompatibility – Liposomes, polymers, metals, etc – Size 1–100 nm	– Molecular modeling and parametrization – Structural characterization (size, shape, packing, etc) – Nanoparticle-ligand binding (free energy calculations)
Second Nanoparticle / Drug / Functionalization	– Active nanostructures – Functionalization – Stability and interaction with biological systems – Selective or specific target recognition – Improved biocompatibility and pharmacokinetics – Funcional groups: PEG, polymers, peptides, lipids, DNA/RRA, etc	– Molecular modeling of functional groups – Evaluation of density/concentration of coating molecules – Nanoparticle-capping agents interactions – Functionalized nanoparticle-ligand affinities (free energy methods) – All-atom and coarse-grained simulations
Third ligand lack / External stimulus / Nanoparticle	– Smart nanostructures – Responsive to external stimuli – Pharmacologically active – Dynamic and multiple functions – Formation of nanostructure aggregates	– Molecular dynamic simulations under different conditions of temperature, pressure, ion concentrations, pH, and force fields – Application of reactive force fields – Surface adsorption energy of nanomaterials – Analysis of nanoparticle-protein corona complexes

Figures were generated with Micelle Maker [27] and adapted from Chu et al. [28] and Ding et al. [29]

well as those mentioned in Table 4.2 to describe classical nanoparticles, could also be used to investigate nanoparticle–nanoparticle interactions.

Overall, progresses in nanotechnology, material engineering, and computer simulation methods offer an unprecedented opportunity to study the molecular structure, targeting and delivery mechanisms, response to stimuli, and multiple functions of nanomaterials. Excellent agreements between experimental and computational data support the use of *in silico* models to assist in the design of nanostructured complexes for biomedical applications. Consequently, molecular computer-aided design software and MD simulation packages, based on semi-empirical and *ab initio* methods,

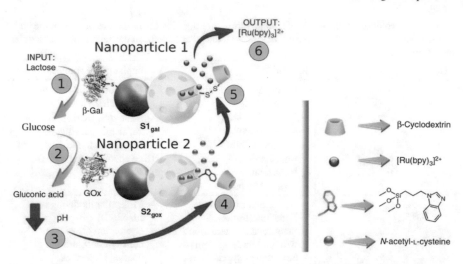

Fig. 4.2 Representation of the interactive communication process between two nanoparticles. Recently, Llopis-Lorente et al. [31] revealed chemical communication between Janus Au-mesoporous silica nanoparticles conjugated with the enzymes β-galactosidase (β-Gal) and glucose oxidase (GOx). A signaling cascade was observed when (1) entering lactose, which is hydrolyzed by β-Gal into galactose and glucose. (2) Glucose is then hydrolyzed into gluconic acid by Gox, (3) provoking a local drop in pH and activating S2$_{gox}$. (4) S2$_{gox}$ activation results in the delivery of entrapped N-acetyl-L-cysteine, which (5) subsequently actives S1$_{gal}$ and induces the rupture of disulfide linkages with (Ru(bpy)$_3$)$^{2+}$. Remarkably, (6) the (Ru(bpy)$_3$)$^{2+}$ delivery only occurs when the two nanoparticles communicate. This phenomenon opens a number of opportunities to develop complex nanoscale systems capable of sharing information and cooperating. Figure adapted from Llopis-Lorente et al. [31]

constitute power and effective strategies to identify biocompatible materials that can be used in detection and treatment of diseases.

4.2 *De novo* Modeling of Nanostructures

Nanomaterials today are based on a wide range of carbon allotropes, metals, ceramics, and synthetic polymers which may be surface-modified with a wide variety of functional groups. As previously discussed, the design of new and improved nanostructures for biomedical uses has become an important challenge. Several studies based on computational techniques report that designing nanostructures with these computational techniques saves time and resources, and the best candidates can then subsequently be tested in biological environments [8, 9, 32]. In computational studies, molecular modeling techniques, force field parameters, initial setups, and systems composition are important factors influencing the outcome of *in silico* simulations used to characterize nanostructure complexes. Thus, to reach reliable predictions, realistic models that exhibit optimal agreements between experimental observations and theoretical simulations are required.

The use of metal-based nanomaterials for biomedical applications has been extensively studied by computational approaches. Particularly, graphene has become an auspicious nanoplatform for uncountable biomedical applications, encompassing areas such as biosensors, absorption of enzymes, drug/gene delivery, cell and tumor imaging, and cancer photothermal therapy [33]. Applications such as these have increased the interest in the manufacturing of graphene and its derivatives, especially graphene nanosheets, ribbons, oxide, and reduced graphene oxide. Recently, graphene (in different oxidations states) has emerged as a promising biocompatible element, due to its ability to interact with a "corona" of adsorbed biomolecules on its surface [34] (Fig. 4.3). The interactions of the graphene protein corona condition the rate and pathway of cellular uptake, physiological distribution, potential routes of excretion, and toxicology of this type of nanoformulation. It has been shown that surface interactions of graphene oxide, as well as the hydration patterns on the surface may generate stabilization/destabilization of several protein structural elements, hauling changes in its biological activity [34, 35]. Thus, materials such as graphene are the central focus in leading of research of potential biomedical applications in nanotechnology.

Molecular dynamics simulations using atomistic models and explicit solvents have been employed to characterize interactions between several nanomaterials and

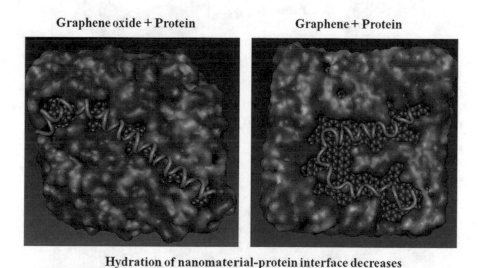

Graphene oxide + Protein   **Graphene + Protein**

Hydration of nanomaterial-protein interface decreases

Helical conformation of protein decreases

Fig. 4.3 Hydration patterns of final graphene-based nanomaterials–protein complexes. The graphene oxide–protein interface is highly hydrated with small areas showing dewetting, whereas the graphene–protein interface is highly dehydrated. Water molecules are shown in surface representation, nanomaterials in spherical representation with gray color, and the protein is highlighted with green color. Figure adapted from Baweja et al. [34]

biomolecules. As an example, molecular dynamics has provided an understanding of the peptide adsorption on metals [36–40] and carbon nanotubes [41]. Carrying agents for drugs and nucleic acids such as dendrimers [24, 42–44] and polymer-based nanoparticles [45, 46] have also been studied through MD simulations. Thus, the use of nanomaterials for biomedical applications has generated the need for designing realistic computational models to study the behavior of nanoparticles at the molecular level. Large carbon nanotubes represent a particularly simple case of study given their chemical structure and well-known surface topography. Classical models of carbon nanotubes and their derivatives have been used as adsorbate on MD simulations and have shown good correlation with experimental data [35]. Additionally, functionalized graphene surfaces have been developed to efficiently reproduce the adsorption free energy, as well as the organization of adsorbates on different graphene surface models. Recently, Singam et al. used molecular dynamics simulations, supplemented by analytical chemistry to explore the adsorption thermodynamics of a diverse set of aromatic compounds on graphene materials, elucidating the effects of the solvent, surface coverage, surface curvature, defects, and functionalization by hydroxy groups [47].

Figure 4.4 shows an example of atomistic models of graphene nanotubes and curved surfaces. Interestingly, the concave curvature has been associated with an

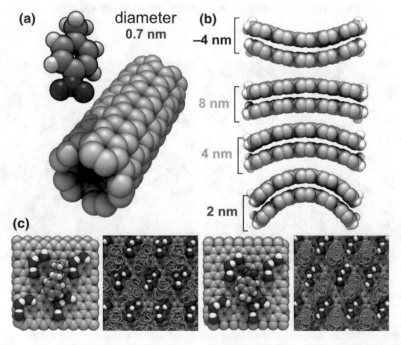

Fig. 4.4 Graphene models at the atomic level. **a** Model of a graphene nanotube having a diameter of 0.7 nm. **b** Graphene surfaces models for different curvatures having diameters of 2.0, 4.0, 8.0, and −4.0 nm. **c** Solute adsorbed organization on hydroxylated graphene models. Adapted from Singam et al. [47]

increase in the adsorption affinity, more favorable enthalpy, and greater contact area for external molecules, whereas convex curvature reduces both adsorption enthalpy and contact area [47]. Comer et al. showed a method to decompose the adsorption free energies into entropic and enthalpic components and found that different classes of compounds—such as phenols, benzoates, and alkylbenzenes—can easily be distinguished by the relative contributions of entropy and enthalpy to their adsorption free energies [47].

The need to produce more reliable computational approaches to reach better agreement with experimental data has driven the development of new and more realistic models of metal nanoparticles. Polarizable models of graphene and carbon nanotubes have been implemented in the force field GRAPPA, which is derived from extensive plane-wave DFT calculations [48]. Polarizable graphene surface models have shown to properly capture the relevant physics and chemistry of complex bio-interfaces, allowing interoperability with known biological force fields.

The use of reactive force fields to model graphene surfaces containing some defects and irregularities has recently been demonstrated. To model these defects, Singam et al. heated and cooled a graphene pristine model using ReaxFF simulations. Figure 4.5 shows the resulting surfaces; rather than being composed of only six-member rings, the defective graphene patch shows 10 five-member rings, 8 seven-member rings, and 1 eight-member ring. This structure agrees with graphene surfaces experimentally observed by electron microscopy [49]. Free energy calculations using defective graphene models showed that the concave curvature topography (see Fig. 4.4) might have higher affinity for adsorbates than convex curvature regions [47].

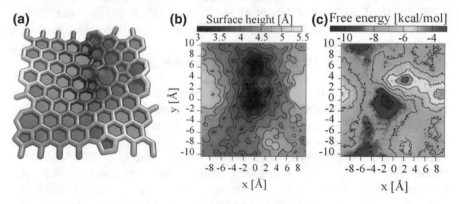

Fig. 4.5 Effect of graphene defects on adsorption thermodynamics of 4-nitrotoluene. **a** Topology and topography of the defective graphene structure. Surface height is indicated by the darkness of the bonds, with depressions shown in black. **b** Surface height as a function of lateral position over the surface. The height was measured along the z axis from the center of mass of the graphene atoms using a probe atom of radius 3 Å. **c** Minimum free energy for 4-nitrotoluene adsorption over this surface determined from a three-dimensional free energy calculation. Adapted from Singam et al. [47]

Another example of nanomaterials used for biomedical applications is gold. Because of its ease of synthesis, functionalization, shape control, biocompatibility, and tunable optical properties, gold has become a nanoparticle widely used in biomedical science. Potential clinical applications of gold include wound treatment [50], drug delivery [51, 52], diagnostic images [53], and various others therapeutic nanotechnologies [53]. Computational methods, such as coarse-grained Monte Carlo simulations, density functional theory (DFT) calculations, and MD simulations, have been extensively employed to reveal the atomistic basis of protein–Au interactions [54]. All-atom simulations have shown the best compromise between accuracy and computational feasibility, emerging as the strongest tool for describing the interaction between gold and organic molecules [55], amino acids [56–58], peptides [59], polymers [60], and proteins [56].

The newest models of gold nanoparticles have been oriented to the development of polarizable surfaces able to better represent the distinctive properties of metal structures and enhance the description of gold–biomolecules interactions. In 2013, Hughes et al. introduced the polarizable force field GolP [56], which is focused purely on describing gold interfaces. This force field captures polarization via the rigid-rod dipole approach [61] and uses virtual interaction sites to ensure a top-site adsorption [62]. In the rigid-rod model, the gold atoms are displayed as dipoles spheres, bound directly with a fixed length distance [56, 61]. These dipoles can freely rotate and generate the polarization effect of the metal. GolP [56] is compatible with the OPLS-AA force field [63], which allows to reproduce interactions between proteins and gold surfaces. However, GolP has shown to predict incorrect adsorption geometries of water molecules on explicit solvent simulations; in particular, hydrogen atoms are oriented toward the nanoparticle surface, which disagrees with quantum mechanical calculations [64]. A CHARMM [65] compatible GolP-CHARMM force field version [66, 67] solved the adsorption geometry of water, and added corrections associated with the adoption of small organic molecules to Au(111) and Au(100) surfaces. Recently, Perfilieva et al. [68] reported two polarizable gold models of 6 and 14 nm in the form of truncated octahedron nanoparticles. The models contain both Au (111) and Au (100) surfaces and reproduce the polarization effects based on the previously described GolP-CHARMM force field [66, 67]. Figure 4.6a shows the core basis of an octahedral polarizable gold model, which was replicated to build the final 6 and 14 nm nanoparticles. The adsorption of the classical capping agent citrate was evaluated against these models, revealing the formation of a stable "corona" of citrates after ≈20 ns of MD simulation (see Fig. 4.6b). Additionally, a non-equal distribution of citrate molecules between the Au (111) and Au (100) surfaces was observed (see Fig. 4.6c), supporting the preference of citrate for Au (111) as it was previously proposed by Park and coauthors [55].

A replica of GolP-CHARMM was constructed for silver surfaces, namely—the AgP-CHARMM force field [69]. This force field was designed for the simulation of biomolecules in aqueous environments on Ag (111) and Ag (100) surfaces. The two metal surfaces are quite similar changing only the atomic weight, the lattice distance between the center of mass of each metal atom and the atomic charge. In

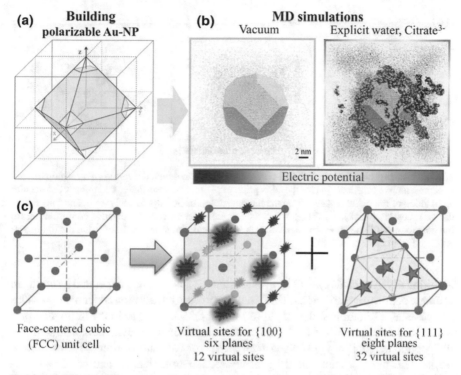

(a) Building polarizable Au-NP

(b) MD simulations

Vacuum   Explicit water, Citrate³⁻

2 nm

Electric potential

(c)

Face-centered cubic (FCC) unit cell

Virtual sites for {100} six planes 12 virtual sites

Virtual sites for {111} eight planes 32 virtual sites

Fig. 4.6 **a** Geometry of a gold nanoparticle obtained from a cube crystal (thin black lines). Cutting planes {111} (thick black lines) form an octahedron in the center of the cube. **b** Final models of the octahedron gold nanoparticle in (left) vacuum and (right) coated by citrate molecules, surrounded by explicit water environment. **c** Unit cell of the face-centered cubic lattice (FCC) with additional virtual sites for {100} and {111} planes (four sites shown). Real atoms of the lattice are represented by orange balls. Virtual sites are shown as red and blue stars for {100} and {111} planes, respectively. Examples of {100} and {111} planes are yellow. Figure adapted from Perfilieva et al. [68]

this case, water overlay stability was observed, comparing the arrangement (spatial and orientational) of water molecules adsorbed to the surface, which agrees with physicochemical properties previously described.

The application of silver nanoparticles for medical purposes has significantly increased, therefore, the use of computational tools based on polarizable silver nanoparticles has helped to rationalize the design of novel and functional nanomaterials. Several articles have established that the incorporation of sub-nanomolar concentrations of nanoparticles mixed with biometric matrices is sufficient to modulate properties such as elasticity, enzymatic degradation resistance, antimicrobial effects, and electroconductivity [37, 70–77]. Poblete et al. [37] proposed that the combination of the thiol groups in cysteine residues and amines in lysine/arginine residues is critical to provide stable protection for the polarizable silver surface along its synthesis. This study denotes the use of the Cys-Leu-Lys sequence as an anchor motif for the adsorption and stabilization of small peptides to Ag nanoparticles [37].

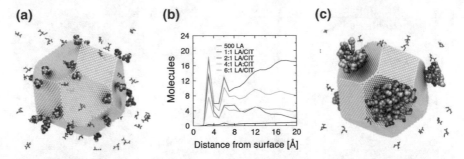

Fig. 4.7 Adsorption of lipoic acid and citrate molecules on octahedral 6-nm silver nanoparticles. **a** Representative conformation for the adsorption of 500 citrate molecules. **b** Number of molecules as function of the radial distance from the nanoparticle, for 500 lipoic acids (red), and lipoic acid: citrate ratios 1:1 (black), 2:1 (blue), 4:1 (orange), 6:1 (green). **c** Representative conformation for the adsorption of lipoic acid (van der Waals spheres) and citrate in a ratio of 6:1. This figure was extracted from Goel et al. [78]

Additionally, Ahumada et al. [75] revealed that the stabilization generated by collagen mimetic peptides containing the CLK motif as anchoring arms on silver nanoparticles increases as a function of the length of the peptide, as well as in the biological performance of the composite [75]. On the other hand, Jacques et al. demonstrated that the peptide length is not the only critical factor for the stabilization of the nanocomposite, but, more crucially, is the supramolecular organization and packing of the peptide once is bound to the nanosurface [76]. Recently, Goel et al. [78] showed that it is possible to estimate the surface area occupied by thiol-containing molecules as function of the nanoparticle concentration. These analyses were obtained by combining experimental approaches (isothermal titration calorimetry, ITC) and computational techniques [73, 78]. The total silver surface coated with small thiol-containing molecules (e.g., lipoic acid), as well as the rationally engineered cysteine-containing peptides (CLK-GP-Hyp-GP-Hyp-GP (peptide 1) and CLK-K(-KLC)-GP-Hyp-GP-Hyp-GP (peptide 2)) was estimated (see Fig. 4.7a and c, respectively). These data propose an approximation of the total surface area available for binding and the differences in the concentration of external molecules as function of the distance to the surface (see Fig. 4.5b). It was also demonstrated that supramolecular arrangements of molecules in the near proximity of the nanosilver surface strongly affects the affinity of thiol-containing molecules [78].

Overall, the *de novo* design of metal nanosurfaces employing polarizable force fields has made the study of large molecular systems possible, which can be simulated by fairly reachable periods of time (up to microseconds). These types of polarizable metal force fields allow the exploration of the adsorption of external molecules in order to design and develop different nanomaterials for biomedical applications. As a future challenge, the effect of polarization should also be explored on biomolecules. In this regard, polarization effects for biomolecules have been introduced through the Drude force field [79], in which a dummy electron is added to a positively charged core, generating a dipole with two opposite charges bound by a harmonic bond. These types of models can be combined with standard force fields to simulate biological molecules and nanoparticles on classical MD software without the need of additional parameters.

4.3 Biomolecules Associated with Nanostructures

Nanomaterials are similar in size when compared to internal cellular components, offering diverse opportunities to influence signaling pathways and biological processes. Nano-engineering materials have promoted the development of numerous systems with multiple applications on nano-biotechnology and nanomedicine [80]. Structural and physicochemical properties such as atomic composition, shape, surface, hydrophobicity, and charge make each nanostructure unique. These properties are keys for achieving the expected function of the nanomaterial; however, size mainly determines the interaction with different biomolecules such as proteins, lipids, sugars, DNA/RNA, and others.

Physicochemical aspects describing the properties of nanostructures in biological environments have been studied through experimental and theoretical approaches. While experimental methods can have limitations when studying the molecular interactions at the nanoscale, theoretical/computational studies have gained attention due to the massive development of molecular models and methods to describe biological nanocomplexes. Computational methods used include electronic-based calculations, quantum mechanics/molecular mechanics approaches (QM/MM), all-atom molecular dynamics (MD), Monte Carlo (MC) simulations, and coarse-grained (CG) methods. Such methods are used to better understand the non-bonding interactions comprising the complexation of nanomaterials and biomolecules and gain information about physical processes occurring in timescales of nano- and microseconds.

Electronic-based methods are highly accurate for calculating binding energies and estimating chemical barriers in reactions. However, these methods are limited to small systems (~100 atoms) and simulation timescales in order of picoseconds, which makes the simulation of nanomaterials in contact with biological molecules difficult, especially considering that the number of atoms per complex reaches the millions [81]. In this regard, the most used techniques for studying the structure–activity relationships and dynamical processes in biological systems are the classical MD approaches. All-atom MD simulations are based in Newton's motion equations for the description of atomic interactions, incorporating forces from many-body interatomic potentials obtained by mathematical approximations in combination with empirical parameters known as force fields. Each force field describes internal bond stretching between atoms, bending and rotation angles, and non-bonded connections such as coulombic and van der Waals interactions. Although there are multiple force fields for a wide range of organic/inorganic nanostructures and biological molecules, their use is limited by the available parameters for each system.

Several studies based on MD simulations have described the interaction of different nanostructures and biomolecules, showing mostly interactions of nanomaterials with proteins, peptides, biological membranes, and DNA/RNA. In general, simulations of these complexes have achieved reasonable descriptions of the mechanisms implicated on the adsorption of biomolecules, physicochemical properties underlying biological processes, binding free energy estimations, and multiple structural descriptions. The effects of nanomaterials under different biological conditions have been also evaluated. Yarovsky et al. [82] used all-atom MD and MC simulations to get perspectives about the interactions between peptides and a SiO_2 surface, providing important insights into the underlying principles of peptide folding and peptide—surface interactions. Noon et al. [83] studied the interaction between antibodies and fullerene nanoparticles through MD simulations, revealing that π-stacking interactions between the side chains of aromatic residues and carbon rings are fundamental for the stabilization of the complex. The insertion of a DNA molecule into a carbon nanotube in a water solute environment was also studied by Gao et al. [84], where van der Waals and hydrophobic forces were found to be essential in the insertion process (Fig. 4.8). These results suggest potential applications of carbon nanotubes in molecular sensors, electronic DNA sequencing, and gene delivery systems. Raffaini and Ganazzoli evaluated the adsorption of fibronectin type 1 module into a hydrophobic graphite surface by atomistic simulations. At initial stages, they detected local adsorption rearrangements of the protein on the surface, whereas at the end of the simulations, an increase in the number of residues in contact with the surface was observed, which means a higher binding energy [85]. Similar computational works have included simulations of nanoparticles with nucleic acids [84, 86, 87], lipids [88–90], amino acids, peptides, and proteins [91, 92]. Large-size complexes have been also described by Ngo et al., where supercrystals of DNA-functionalized gold nanoparticles containing 2.77 and 5.05 million atoms were analyzed by MD simulations [93].

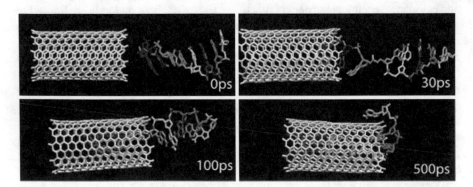

Fig. 4.8 Simulation snapshots of a DNA oligonucleotide (8 adenine bases) interacting with a (10,10) carbon nanotube at 0, 30, 100, and 500 ps of MD simulation. This image was extracted from Gao et al. [84]

Because of limitations in computational resources, early MD simulations were reduced to small molecular models of biomolecule/nanomaterial interfaces and were restricted to timescales of a few nanoseconds with a limited number of atoms. Currently, the incorporation of GPU video cards and the optimization of MD packages have made the structural characterizations of diverse nanomaterials combined with complex biological structures, such as large proteins, feasible. Recent studies of nanomaterials associated with biomolecules have included oxides, minerals in different forms (titanium dioxides, silicon dioxides, etc.), organic polymers, nanotubes, and others. Among the most favorable inorganic nanomaterials for medical use highlight metallic nanostructures [7] such as surface-engineered silver nanoparticles (AgNPs) and gold nanoparticles (AuNPs). These have been analyzed through MD simulations contemplating from small models (surfaces) to large-scale models with different shapes (spherical, pseudospherical, cubic) and sizes. In 2015, Ramezani et al. reported new insights about the interaction between homotripeptides and gold nanoparticles, showing the effects of flexibility, size, and stability of the peptides on the surface [94]. On the other hand, functionalized AgNPs have shown to improve the stability and biocompatibility of nanoparticles, enhancing their application as small-molecule delivery systems and antimicrobial agents [37, 75]. In addition, specific domains of peptides have been reported for capping/stabilizing AgNPs, which has been demonstrated by MD simulations and biological assessments.

To explore the binding affinity of biomolecules on nanostructures, a replica exchange MD (REMD) method has been developed by Sugita et al. [95], which is based on multiple and independent MD simulations under different conditions (temperatures or Hamiltonians). Using this method, it is possible to overcome high energy barriers and sample a large set of configurations for nanoparticle: biomolecule complexes. Temperature-based REMD (T-REMD) has been used to improve the sampling of peptide–surface interactions in several studies [96–99], whereas a variation of REMD known as replica exchange with solute tempering (REST) [100] was developed for enhancing poor scaling in large-size systems [101]. Hosseinzadeh et al. studied the structural changes of the insulin protein interacting with ZnO nanoparticles (Fig. 4.9) by REMD simulations, showing that the polar and charged amino acids of insulin are keys for its association with the nanoparticle [99].

While sampling in MD is important to get information about the structural arrangements and to explore different configurations of the systems, free energy calculations are used for determining binding affinities of biomolecules and surfaces. In this regard, kinetics and transition free energies in adsorption processes are critical for linking theoretical and experimental results. Multiple methods are currently available for estimating binding free energy, which varies in accuracy and complexity. Most of these approaches combine molecular dynamics and Monte Carlo simulations for sampling. The molecular mechanics generalized Born surface area (MM-GBSA) and Poisson–Boltzmann surface area (MM-PBSA) methods are defined as end-point approaches, where only the reference unbound and the final bound states are sampled to obtain the free energy difference. These methods are commonly used for protein–ligand interactions; however, recent studies have incorporated MM-GBSA/PBSA for evaluating peptide-nanostructure interactions [102, 103]. Others alternatives include pathway methods, which are more sophisticated and involve high computational

Fig. 4.9 Interaction site of insulin with ZnO nanoparticles by REMD simulations. This figure was extracted from Hosseinzadeh et al. [99]

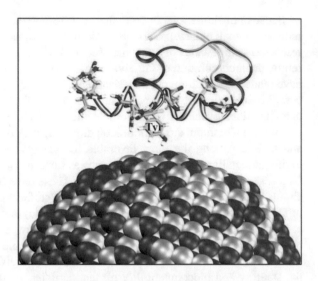

effort to calculate binding free energies using different routes such as: (a) probability density, (b) free energy perturbation, (c) thermodynamic integration, and (d) nonequilibrium work approaches [104]. Among the most used methods in the study of the interaction between biomolecules and nanomaterials have been described: umbrella sampling [105, 106], metadynamics [107], steered molecular dynamics (SMD) [108], and adaptive biasing force (ABF) [109]. These methods have been employed in multiple studies including nanoparticle adsorption into lipid membranes [110–112], peptide–surface interactions [37, 76, 113–118], and adsorption of small molecules into organic/inorganic surfaces [35, 47].

More complex nanosystems have been recently studied by atomistic MD simulations. Lai et al. modeled a 5–6-nm silver nanoparticle functionalized with lipoproteins (HDL) to evaluate its function as a powerful therapeutic agent for removing excess of cholesterol from arterial plaques. These studies included coarse-grained and all-atom MD simulations supported by X-ray photoelectron spectroscopy and lecithin/cholesterol acyltransferase activation experiments [119]. Likewise, Perfilieva et al. developed a polarizable model for AuNPs (6 and 14 nm of diameter) interacting with sodium citrate. They found that polarization effects in the metal determine the mechanisms of interaction of the gold nanoparticle and citrate ions [68].

4.4 Concluding Remarks

In conclusion, computational tools have helped to characterize several nanostructured systems; however, due to the multiple physical and chemical contributions that regulate the association of biomolecules with inorganic/organics nanomaterials, the development of more sophisticated computational models and software is required for studying nanostructures. Future perspectives in this field include the improvement of the force fields and MD simulation programs, which will capture more realistic properties currently not included in conventional MD platforms, such as chemical-bond formation, charge distribution during time, among other effects [120–122]. Moreover, the improvement of highly precise computational methodologies, plus the constant growth in the computational power available, will advance the modeling of nanoparticles that can be used in a wide range of applications. With the same spirit, the improvement in the computational methods complemented with experiments will lead the development of new technologies with powerful biomedical applications.

Acknowledgements Authors thank FONDECYT grant no. 1171155 and 11170223 as well as Millennium Nucleus of Ion Channel-Associated Diseases (MiNICAD); a Millennium Nucleus supported by the Iniciativa Cientifica Milenio of the Ministry of Economy, Development and Tourism (Chile).

Disclosures All authors have read and approved this final version.

References

1. Duncan R. The dawning era of polymer therapeutics. Nat Rev Drug Discov. 2003;2(5):347–60.
2. Gao X, Cui Y, Levenson RM, Chung LW, Nie S. In vivo cancer targeting and imaging with semiconductor quantum dots. Nat Biotechnol. 2004;22(8):969–76.
3. Huh Y-M, Jun Y-w, Song H-T, Kim S, Choi J-s, Lee J-H, Yoon S, Kim K-S, Shin J-S, Suh J-S and others. In vivo magnetic resonance detection of cancer by using multifunctional magnetic nanocrystals. J Am Chem Soc. 2005;127(35):12387–91.
4. Sengupta S, Eavarone D, Capila I, Zhao G, Watson N, Kiziltepe T, Sasisekharan R. Temporal targeting of tumour cells and neovasculature with a nanoscale delivery system. Nature. 2005;436(7050):568–72.
5. Popovic N, Brundin P. Therapeutic potential of controlled drug delivery systems in neurodegenerative diseases. Int J Pharm. 2006;314(2):120–6.
6. Khademhosseini A, Langer R. A decade of progress in tissue engineering. Nat Protoc. 2016;11:1775.
7. Verma S, Domb AJ, Kumar N. Nanomaterials for regenerative medicine. Nanomedicine. 2010;6(1):157–81.
8. Merkle RC. Computational nanotechnology. Nanotechnology. 1991;2:134–41.
9. Tetley TD. Health effects of nanomaterials. Biochem Soc Trans. 2007;35(Pt 3):527–31.
10. Saini B, Srivastava S. Nanoinformatics: predicting toxicity using computational modeling. Comput Intell Big Data Anal. 2018. pp. 65–73.
11. Albanese A, Tang PS, Chan WCW. The effect of nanoparticle size, shape, and surface chemistry on biological systems. Ann Rev Biomed Eng. 2012;14:1–16.
12. Bora T, Dousse A, Sharma K, Sarma K, Baev A, Hornyak GL, Dasgupta G. Modeling nanomaterial physical properties: theory and simulation. Int J Smart Nano Mater. 2019;10(2):116–43.
13. Lamon L, Asturiol D, Vilchez A, Ruperez-Illescas R, Cabellos J, Richarz A, Worth A. Computational models for the assessment of manufactured nanomaterials: development of model reporting standards and mapping of the model landscape. Comput Toxicol. 2019;9:143–51.
14. Petros RA, DeSimone JM. Strategies in the design of nanoparticles for therapeutic applications. Nat Rev Drug Discov. 2010;9(8):615–27.

15. Barenholz Y. Doxil®–the first FDA-approved nano-drug: lessons learned. J Control Release. 2012;160(2):117–34.
16. Yacoub Tyrone J, Reddy Allam S, Szleifer I. Structural effects and translocation of doxorubicin in a DPPC/Chol bilayer: the role of cholesterol. Biophys J. 2011;101(2):378–85.
17. Xiang T-X, Jiang Z-Q, Song L, Anderson BD. Molecular dynamics simulations and experimental studies of binding and mobility of 7-tert-butyldimethylsilyl-10-hydroxycamptothecin and its 20(S)-4-aminobutyrate ester in DMPC membranes. Mol Pharm. 2006;3(5):589–600.
18. Xiang T-X, Anderson BD. Liposomal drug transport: a molecular perspective from molecular dynamics simulations in lipid bilayers. Adv Drug Deliv Rev. 2006;58(12–13):1357–78.
19. Pelaz B, del Pino P, Maffre P, Hartmann R, Gallego M, Rivera-Fernández S, de la Fuente JM, Nienhaus GU, Parak WJ. Surface functionalization of nanoparticles with polyethylene glycol: effects on protein adsorption and cellular uptake. ACS Nano. 2015;9(7):6996–7008.
20. Shen Z, Loe DT, Awino JK, Kröger M, Rouge JL, Li Y. Self-assembly of core-polyethylene glycol-lipid shell (CPLS) nanoparticles and their potential as drug delivery vehicles. Nanoscale. 2016;8(31):14821–35.
21. Haume K, Mason NJ, Solov'yov AV. Modeling of nanoparticle coatings for medical applications. Eur Phys J D. 2016;70(9):181.
22. Marasini R, Pitchaimani A, Nguyen TDT, Comer J, Aryal S. The influence of polyethylene glycol passivation on the surface plasmon resonance induced photothermal properties of gold nanorods. Nanoscale. 2018;10(28):13684–93.
23. Karimi M, Ghasemi A, Sahandi Zangabad P, Rahighi R, Moosavi Basri SM, Mirshekari H, Amiri M, Shafaei Pishabad Z, Aslani A, Bozorgomid M and others. Smart micro/nanoparticles in stimulus-responsive drug/gene delivery systems. Chem Soc Rev. 2016;45(5):1457–1501.
24. Vergara-Jaque A, Comer J, Monsalve L, González-Nilo FD, Sandoval C. Computationally Efficient methodology for atomic-level characterization of dendrimer-drug complexes: a comparison of amine- and acetyl-terminated PAMAM. J Phys Chem B. 2013;117(22):6801–13.
25. Khalilov U, Bogaerts A, Neyts EC. Atomic scale simulation of carbon nanotube nucleation from hydrocarbon precursors. Nat Commun. 2015;6:10306.
26. Shimizu K, Nakamura H, Watano S. MD simulation study of direct permeation of a nanoparticle across the cell membrane under an external electric field. Nanoscale. 2016;8(23):11897–906.
27. Krüger DM, Kamerlin SCL. Micelle maker: an online tool for generating equilibrated micelles as direct input for molecular dynamics simulations. ACS Omega. 2017;2(8):4524–30.
28. Chu Z, Han Y, Bian T, De S, Král P, Klajn R. Supramolecular control of azobenzene switching on nanoparticles. J Am Chem Soc. 2019;141(5):1949–60.
29. Ding H-m, Ma Y-q. Controlling cellular uptake of nanoparticles with pH-sensitive polymers. Sci Rep. 2013;3:2804.
30. von Maltzahn G, Park J-H, Lin KY, Singh N, Schwöppe C, Mesters R, Berdel WE, Ruoslahti E, Sailor MJ, Bhatia SN. Nanoparticles that communicate in vivo to amplify tumour targeting. Nat Mater. 2011;10(7):545–52.
31. Llopis-Lorente A, Díez P, Sánchez A, Marcos MD, Sancenón F, Martínez-Ruiz P, Villalonga R, Martínez-Máñez R. Interactive models of communication at the nanoscale using nanoparticles that talk to one another. Nat Comm. 2017;8:15511.
32. Saini B, Srivastava S. Nanoinformatics: Predicting Toxicity Using Computational Modeling. In: Satyanarayana C, Rao KN, Bush RG, editors. Computational intelligence and big data analytics: applications in bioinformatics. Singapore: Springer; 2019. p. 65–73.
33. Hu W, Peng C, Lv M, Li X, Zhang Y, Chen N, Fan C, Huang Q. Protein corona-mediated mitigation of cytotoxicity of Graphene Oxide. ACS Nano. 2011;5(5):3693–700.
34. Baweja L, Balamurugan K, Subramanian V, Dhawan A. Hydration patterns of Graphene-Based Nanomaterials (GBNMs) play a major role in the stability of a helical protein: a molecular dynamics simulation study. Langmuir. 2013;29(46):14230–8.
35. Comer J, Chen R, Poblete H, Vergara-Jaque A, Riviere JE. Predicting adsorption affinities of small molecules on carbon nanotubes using molecular dynamics simulation. ACS Nano. 2015;9(12):11761–74.

36. Slocik JM, Naik RR. Probing peptide-nanomaterial interactions. Chem Soc Rev. 2010;39(9):3454–63.
37. Poblete H, Agarwal A, Thomas SS, Bohne C, Ravichandran R, Phopase J, Comer J, Alarcon EI. New Insights into peptide-silver nanoparticle interaction: deciphering the role of cysteine and lysine in the peptide sequence. Langmuir. 2016;32(1):265–73.
38. Heinz H, Farmer BL, Pandey RB, Slocik JM, Patnaik SS, Pachter R, Naik RR. Nature of molecular interactions of Peptides with Gold, Palladium, and Pd−Au bimetal surfaces in aqueous solution. J Am Chem Soc. 2009;131(28):9704–14.
39. Coppage R, Slocik JM, Briggs BD, Frenkel AI, Heinz H, Naik RR, Knecht MR. Crystallographic recognition controls peptide binding for bio-based nanomaterials. J Am Chem Soc. 2011;133(32):12346–9.
40. Heinz H, Lin T-J, Kishore Mishra R, Emami FS. Thermodynamically consistent force fields for the assembly of inorganic, organic, and biological nanostructures: the INTERFACE force field. Langmuir. 2013;29(6):1754–65.
41. Katoch J, Kim SN, Kuang Z, Farmer BL, Naik RR, Tatulian SA, Ishigami M. Structure of a peptide adsorbed on Graphene and Graphite. Nano Lett. 2012;12(5):2342–6.
42. Caballero J, Poblete H, Navarro C, Alzate-Morales JH. Association of nicotinic acid with a poly(amidoamine) dendrimer studied by molecular dynamics simulations. J Mol Graph Model. 2013;39:71–8.
43. Maingi V, Kumar MVS, Maiti PK. PAMAM dendrimer-drug interactions: Effect of pH on the binding and release pattern. J Phys Chem B. 2012;116(14):4370–6.
44. Carrasco-Sánchez V, Vergara-Jaque A, Zúñiga M, Comer J, John A, Nachtigall FM, Valdes O, Duran-Lara EF, Sandoval C, Santos LS. In situ and in silico evaluation of amine- and folate-terminated dendrimers as nanocarriers of anesthetics. Eur J Med Chem. 2014;73:250–7.
45. Sun C, Tang T, Uludag H. A molecular dynamics simulation study on the effect of lipid substitution on polyethylenimine mediated siRNA complexation. Biomaterials. 2013;34(11):2822–33.
46. Vilos C, Morales FA, Solar PA, Herrera NS, Gonzalez-Nilo FD, Aguayo DA, Mendoza HL, Comer J, Bravo ML, Gonzalez PA and others. Paclitaxel-PHBV nanoparticles and their toxicity to endometrial and primary ovarian cancer cells. Biomaterials. 2013;34(16):4098–4108.
47. Azhagiya Singam ER, Zhang Y, Magnin G, Miranda-Carvajal I, Coates L, Thakkar R, Poblete H, Comer J. Thermodynamics of adsorption on graphenic surfaces from aqueous solution. J Chem Theory Comput. 2019;15(2):1302–16.
48. Hughes ZE, Tomásio SM, Walsh TR. Efficient simulations of the aqueous bio-interface of graphitic nanostructures with a polarisable model. Nanoscale. 2014;6(10):5438–48.
49. Robertson AW, Warner JH. Atomic resolution imaging of graphene by transmission electron microscopy. Nanoscale. 2013;5(10):4079–93.
50. Li Y, Tian Y, Zheng W, Feng Y, Huang R, Shao J, Tang R, Wang P, Jia Y, Zhang J and others. Composites of bacterial cellulose and small molecule-decorated gold nanoparticles for treating gram-negative bacteria-infected wounds. Small 2017;13(27):1700130.
51. Lei Y, Tang L, Xie Y, Xianyu Y, Zhang L, Wang P, Hamada Y, Jiang K, Zheng W, Jiang X. Gold nanoclusters-assisted delivery of NGF siRNA for effective treatment of pancreatic cancer. Nat Commun. 2017;8:15130.
52. Deyev S, Proshkina G, Ryabova A, Tavanti F, Menziani MC, Eidelshtein G, Avishai G, Kotlyar A. Synthesis, characterization, and selective delivery of DARPin–Gold nanoparticle conjugates to cancer cells. Bioconjugate Chem. 2017;28(10):2569–74.
53. Lim E-K, Kim T, Paik S, Haam S, Huh Y-M, Lee K. Nanomaterials for theranostics: recent advances and future challenges. Chem Rev. 2015;115(1):327–94.
54. Charchar P, Christofferson AJ, Todorova N, Yarovsky I. Understanding and designing the gold-bio interface: insights from simulations. Small. 2016;12(18):2395–418.
55. Park J-W, Shumaker-Parry JS. Structural study of citrate layers on gold nanoparticles: role of intermolecular interactions in stabilizing nanoparticles. J Am Chem Soc. 2014;136(5):1907–21.

56. Iori F, Di Felice R, Molinari E, Corni S. GolP: an atomistic force-field to describe the interaction of proteins with Au(111) surfaces in water. J Comput Chem. 2009;30(9):1465–76.

57. Hoefling M, Iori F, Corni S, Gottschalk K-E. Interaction of Amino Acids with the Au(111) surface: adsorption free energies from molecular dynamics simulations. Langmuir. 2010;26(11):8347–51.

58. Hoefling M, Monti S, Corni S, Gottschalk KE. Interaction of β-sheet folds with a gold surface. PLoS ONE. 2011;6(6):e20925.

59. Tang M, Gandhi NS, Burrage K, Gu Y. Interaction of gold nanosurfaces/nanoparticles with collagen-like peptides. Phys Chem Chem Phys. 2019;21(7):3701–11.

60. Camarada MB, Comer J, Poblete H, Azhagiya Singam ER, Marquez-Miranda V, Morales-Verdejo C, Gonzalez-Nilo FD. Experimental and computational characterization of the interaction between gold nanoparticles and polyamidoamine dendrimers. Langmuir. 2018;34(34):10063–72.

61. Iori F, Corni S. Including image charge effects in the molecular dynamics simulations of molecules on metal surfaces. J Comput Chem. 2008;29(10):1656–66.

62. Chen D-L, Al-Saidi WA, Johnson JK. Noble gases on metal surfaces: insights on adsorption site preference. Phys Rev B. 2011;84(24):241405.

63. Jorgensen WL, Maxwell DS, Tirado-Rives J. Development and testing of the OPLS all-atom force field on conformational energetics and properties of organic liquids. J Am Chem Soc. 1996;118(45):11225–36.

64. Cicero G, Calzolari A, Corni S, Catellani A. Anomalous wetting layer at the Au(111) surface. J Phys Chem Lett. 2011;2(20):2582–6.

65. MacKerell AD, Bashford D, Bellott M, Dunbrack RL, Evanseck JD, Field MJ, Fischer S, Gao J, Guo H, Ha S and others. All-atom empirical potential for molecular modeling and dynamics studies of proteins. J Phys Chem B 1998;102(18):3586–3616.

66. Wright LB, Rodger PM, Walsh TR, Corni S. First-principles-based force field for the interaction of proteins with Au(100)(5 × 1): an extension of GolP-CHARMM. J Phys Chem C. 2013;117(46):24292–306.

67. Wright LB, Rodger PM, Corni S, Walsh TR. GolP-CHARMM: first-principles based force fields for the interaction of proteins with Au(111) and Au(100). J Chem Theory Comput. 2013;9(3):1616–30.

68. Perfilieva OA, Pyshnyi DV, Lomzov AA. Molecular dynamics simulation of polarizable gold nanoparticles interacting with Sodium Citrate. J Chem Theory Comput. 2019;15(2):1278–92.

69. Hughes ZE, Wright LB, Walsh TR. Biomolecular adsorption at aqueous silver interfaces: first-principles calculations, polarizable force-field simulations, and comparisons with Gold. Langmuir. 2013;29(43):13217–29.

70. Alarcon EI, Udekwu KI, Noel CW, Gagnon LBP, Taylor PK, Vulesevic B, Simpson MJ, Gkotzis S, Islam MM, Lee C-J and others. Safety and efficacy of composite collagen–silver nanoparticle hydrogels as tissue engineering scaffolds. Nanoscale 2015;7(44):18789–98.

71. Alarcon EI, Udekwu K, Skog M, Pacioni NL, Stamplecoskie KG, González-Béjar M, Polisetti N, Wickham A, Richter-Dahlfors A, Griffith M and others. The biocompatibility and antibacterial properties of collagen-stabilized, photochemically prepared silver nanoparticles. Biomaterials 2012;33(19):4947–56.

72. Hosoyama K, Ahumada M, McTiernan CD, Bejjani J, Variola F, Ruel M, Xu B, Liang W, Suuronen EJ, Alarcon EI. Multi-functional thermo-crosslinkable collagen-metal nanoparticle composites for tissue regeneration: nanosilver vs. nanogold. RSC Adv. 2017;7(75):47704–08.

73. Lazurko C, Ahumada M, Valenzuela-Henríquez F, Alarcon EI. NANoPoLC algorithm for correcting nanoparticle concentration by sample polydispersity. Nanoscale. 2018;10(7):3166–70.

74. Ahumada M, Lissi E, Montagut AM, Valenzuela-Henríquez F, Pacioni NL, Alarcon EI. Association models for binding of molecules to nanostructures. Analyst. 2017;142(12):2067–89.

75. Ahumada M, Jacques E, Andronic C, Comer J, Poblete H, Alarcon EI. Novel specific peptides as superior surface stabilizers for silver nano structures: role of peptide chain length. J Mater Chem B. 2017;5(45):8925–8.

76. Jacques E, Ahumada M, Rector B, Yousefalizadeh G, Galaz-Araya C, Recabarren R, Stample-coskie K, Poblete H, Alarcon EI. Effect of nanosilver surfaces on peptide reactivity towards reactive oxygen species. Nanoscale. 2018;10(34):15911–7.

77. Dvir T, Timko BP, Brigham MD, Naik SR, Karajanagi SS, Levy O, Jin H, Parker KK, Langer R, Kohane DS. Nanowired three-dimensional cardiac patches. Nat Nanotechnol. 2011;6(11):720–5.

78. Goel K, Zuñiga-Bustos M, Lazurko C, Jacques E, Galaz-Araya C, Valenzuela-Henriquez F, Pacioni NL, Couture J-F, Poblete H, Alarcon EI. Nanoparticle concentration vs surface area in the interaction of thiol-containing molecules: toward a rational nanoarchitectural design of hybrid materials. ACS Appl Mater Inter. 2019;11(19):17697–705.

79. Lemkul JA, Huang J, Roux B, MacKerell AD. An empirical polarizable force field based on the classical drude oscillator model: development history and recent applications. Chem Rev. 2016;116(9):4983–5013.

80. Yanamala N, Kagan VE, Shvedova AA. Molecular modeling in structural nano-toxicology: Interactions of nano-particles with nano-machinery of cells. Adv Drug Deliv Rev. 2013;65(15):2070–7.

81. Ozboyaci M, Kokh DB, Corni S, Wade RC. Modeling and simulation of protein–surface interactions: achievements and challenges. Q Rev Biophys 2016;49.

82. Yarovsky I, Hearn MTW, Aguilar MI. Molecular simulation of peptide interactions with an RP-HPLC sorbent. J Phys Chem B. 1997;101(50):10962–70.

83. Noon WH, Kong Y, Ma J. Molecular dynamics analysis of a buckyball-antibody complex. PNAS. 2002;99(Suppl 2):6466–70.

84. Gao H, Kong Y, Cui D, Ozkan CS. Spontaneous insertion of DNA oligonucleotides into carbon nanotubes. Nano Lett. 2003;3(4):471–3.

85. Raffaini G, Ganazzoli F. Molecular dynamics simulation of the adsorption of a fibronectin module on a graphite surface [†]. Langmuir. 2004;20(8):3371–8.

86. Lu G, Maragakis P, Kaxiras E. Carbon Nanotube Interaction with DNA. Nano Lett. 2005;5(5):897–900.

87. Johnson RR, Johnson ATC, Klein ML. Probing the structure of DNA–carbon nanotube hybrids with molecular dynamics. Nano Lett. 2008;8(1):69–75.

88. Bedrov D, Smith GD, Davande H, Li L. Passive transport of C60 fullerenes through a lipid membrane: a molecular dynamics simulation study. J Phys Chem B. 2008;112(7):2078–84.

89. Li Y, Chen X, Gu N. Computational investigation of interaction between nanoparticles and membranes: hydrophobic/hydrophilic effect. J Phys Chem B. 2008;112(51):16647–53.

90. Wang H, Michielssens S, Moors SLC, Ceulemans A. Molecular dynamics study of dipalmi-toylphosphatidylcholine lipid layer self-assembly onto a single-walled carbon nanotube. Nano Res. 2009;2(12):945–54.

91. Chiu C-C, Dieckmann GR, Nielsen SO. Role of peptide–peptide interactions in stabilizing peptide-wrapped single-walled carbon nanotubes: a molecular dynamics study. Peptide Sci 2009;92(3):156–163.

92. Chiu C-C, Dieckmann GR, Nielsen SO. Molecular dynamics study of a nanotube-binding amphiphilic helical peptide at different water/hydrophobic interfaces. J Phys Chem B 2008;112(51):16326–33.

93. Ngo VA, Kalia RK, Nakano A, Vashishta P. Supercrystals of DNA-functionalized gold nanoparticles: a million-atom molecular dynamics simulation study. J Phys Chem C. 2012;116(36):19579–85.

94. Ramezani F, Habibi M, Rafii-Tabar H, Amanlou M. Effect of peptide length on the conjugation to the gold nanoparticle surface: a molecular dynamic study. DARU J Pharm Sci. 2015;23(1):9.

95. Sugita Y, Okamoto Y. Replica-exchange molecular dynamics method for protein folding. Chem Phys Lett. 1999;314(1):141–51.

96. Corni S, Hnilova M, Tamerler C, Sarikaya M. Conformational behavior of genetically-engineered Dodecapeptides as a determinant of binding affinity for Gold. J Phys Chem C. 2013;117(33):16990–7003.

97. Li H, Luo Y, Derreumaux P, Wei G. Carbon nanotube inhibits the formation of β-sheet-rich oligomers of the Alzheimer's Amyloid-β(16-22) peptide. Biophys J. 2011;101(9):2267–76.
98. Liao C, Zhou J. Replica-exchange molecular dynamics simulation of basic fibroblast growth factor adsorption on hydroxyapatite. J Phys Chem B. 2014;118(22):5843–52.
99. Hosseinzadeh G, Maghari A, Farnia SMF, Moosavi-Movahedi AA. Interaction mechanism of insulin with ZnO nanoparticles by replica exchange molecular dynamics simulation. J Biomol Struct Dyn. 2018;36(14):3623–35.
100. Liu P, Kim B, Friesner RA, Berne BJ. Replica exchange with solute tempering: a method for sampling biological systems in explicit water. PNAS. 2005;102(39):13749–54.
101. Tang Z, Palafox-Hernandez JP, Law W-C, Hughes ZE, Swihart MT, Prasad PN, Knecht MR, Walsh TR. Biomolecular recognition principles for bionanocombinatorics: an integrated approach to elucidate enthalpic and entropic factors. ACS Nano. 2013;7(11):9632–46.
102. Song M, Sun Y, Luo Y, Zhu Y, Liu Y, Li H. Exploring the mechanism of inhibition of au nanoparticles on the aggregation of Amyloid-β(16-22) peptides at the atom level by all-atom molecular dynamics. Int J Mol Sci 2018;19(6).
103. Xie L, Luo Y, Lin D, Xi W, Yang X, Wei G. The molecular mechanism of fullerene-inhibited aggregation of Alzheimer's β-amyloid peptide fragment. Nanoscale. 2014;6(16):9752–62.
104. Free Energy Calculations. Theory and applications in Chemistry and Biology. Berlin, Heidelberg: Springer; 2007.
105. Torrie GM, Valleau JP. Nonphysical sampling distributions in Monte Carlo free-energy estimation: Umbrella sampling. J Comput Phys. 1977;23(2):187–99.
106. Torrie GM, Valleau JP. Monte Carlo free energy estimates using non-Boltzmann sampling: Application to the sub-critical Lennard-Jones fluid. Chem Phys Lett. 1974;28(4):578–81.
107. Laio A, Parrinello M. Escaping free-energy minima. PNAS. 2002;99(20):12562–6.
108. Izrailev S, Stepaniants S, Balsera M, Oono Y, Schulten K. Molecular dynamics study of unbinding of the avidin-biotin complex. Biophys J. 1997;72(4):1568–81.
109. Darve E, Pohorille A. Calculating free energies using average force. J Chem Phys. 2001;115(20):9169–83.
110. Lehn RCV, Alexander-Katz A. Energy landscape for the insertion of amphiphilic nanoparticles into lipid membranes: a computational study. PLoS ONE. 2019;14(1):e0209492.
111. Nademi Y, Tang T, Uludağ H. Steered molecular dynamics simulations reveal a self-protecting configuration of nanoparticles during membrane penetration. Nanoscale. 2018;10(37):17671–82.
112. Li Y, Gu N. Thermodynamics of charged nanoparticle adsorption on charge-neutral membranes: a simulation study. J Phys Chem B. 2010;114(8):2749–54.
113. Ferreira AF, Comune M, Rai A, Ferreira L, Simões PN. Atomistic-level investigation of a LL37-conjugated gold nanoparticle by well-tempered metadynamics. J Phys Chem B. 2018;122(35):8359–66.
114. Prakash A, Sprenger KG, Pfaendtner J. Essential slow degrees of freedom in protein-surface simulations: a metadynamics investigation. Biochem Biophys Res Commun. 2018;498(2):274–81.
115. Zhang S, Liu Q, Cheng H, Gao F, Liu C, Teppen BJ. Thermodynamic mechanism and interfacial structure of kaolinite intercalation and surface modification by alkane surfactants with neutral and ionic head groups. J Phys Chem C Nanomater Interf. 2017;121(16):8824–31.
116. Emami FS, Puddu V, Berry RJ, Varshney V, Patwardhan SV, Perry CC, Heinz H. Prediction of specific biomolecule adsorption on silica surfaces as a function of pH and particle size. Chem Mater. 2014;26(19):5725–34.
117. Deighan M, Pfaendtner J. Exhaustively sampling peptide adsorption with metadynamics. Langmuir. 2013;29(25):7999–8009.
118. Boughton AP, Andricioaei I, Chen Z. Surface orientation of Magainin 2: molecular dynamics simulation and sum frequency generation vibrational spectroscopic studies. Langmuir. 2010;26(20):16031–6.
119. Lai C-T, Sun W, Palekar RU, Thaxton CS, Schatz GC. Molecular dynamics simulation and experimental studies of gold nanoparticle templated HDL-like nanoparticles for cholesterol metabolism therapeutics. ACS Appl Mater Interf. 2017;9(2):1247–54.

120. Monti S, Barcaro G, Sementa L, Carravetta V, Ågren H. Dynamics and self-assembly of bio-functionalized gold nanoparticles in solution: Reactive molecular dynamics simulations. Nano Res. 2018;11(4):1757–67.
121. Monti S, Barcaro G, Sementa L, Carravetta V, Ågren H. Characterization of the adsorption dynamics of trisodium citrate on gold in water solution. RSC Adv. 2017;7(78):49655–63.
122. Monti S, Carravetta V, Ågren H. Decoration of gold nanoparticles with cysteine in solution: reactive molecular dynamics simulations. Nanoscale. 2016;8(26):12929–38.

Chapter 5
Nanomaterials Applications in Cartilage Tissue Engineering

Janani Mahendran and Jean-Philippe St-Pierre

Abstract Articular cartilage is the smooth layer of soft tissue that covers our bones and allows for the painless movement of our joints. Because of joint pathologies such as arthritis, cartilage can degrade over time in some individuals, causing them to live with considerable pain and reduced mobility. The high prevalence of arthritis and the absence of a cure for osteoarthritis, its most common form, have fueled sustained efforts to develop tissue engineering and regenerative medicine strategies aimed at regenerating cartilage. Despite a number of clinical advances that elicit cartilage repair, true regeneration remains elusive. Recent years have seen an increased use of nanoscale materials in the development of therapies for joint pathologies. Nanomaterials are comparable in scale to the principal building blocks of cartilage extracellular matrix, namely collagen and proteoglycan aggregates. Similarly, nanoparticles are sufficiently small to allow diffusion through the pores of the dense cartilage extracellular matrix and cell targeting. In this chapter, the organization of cartilage's main building blocks will be reviewed from the nano- to macroscale, and sub-micron particles that participate in cell-cell communication will be highlighted. Efforts to design scaffolds incorporating cell-instructive nanoscale features and to tailor the mechanical properties, or even engineer spatial organization, in scaffolds for cartilage repair using nanomaterials will also be discussed. Finally, key design criteria in nanoparticle synthesis to enable targeted therapeutic delivery will be examined.

5.1 Introduction

Articular cartilage is a load-bearing connective tissue that covers the articulating surfaces of bones in synovial joints. It is characterized by a dense and highly hydrated extracellular matrix (ECM) mainly comprising a collagen type II fibrillar network

J. Mahendran · J.-P. St-Pierre (✉)
Department of Chemical and Biological Engineering, University of Ottawa,
161 Louis-Pasteur, Ottawa, ON K1N 6N5, Canada
e-mail: Jean-Philippe.St-Pierre@uOttawa.ca

© Springer Nature Switzerland AG 2019
E. I. Alarcon and M. Ahumada (eds.), *Nanoengineering Materials for Biomedical Uses*,
https://doi.org/10.1007/978-3-030-31261-9_5

interspersed with proteoglycans, as well as other collagens and non-collagenous proteins [1]. The interplay between factors such as composition, orientation and cross-linking of these biomolecules dictates the hydration of the tissue and its mechanical properties [2]. This complex organization allows the tissue to resist shear and tensile forces applied to the joint during articulation while facilitating the absorption and distribution of compression forces transmitted through the joint [3]. Chondrocytes—the resident cells in articular cartilage—are sparsely distributed throughout the tissue, where they remodel the ECM in response to the changing biomechanical environment and help maintain tissue integrity [4].

Under physiological conditions, articular cartilage homeostasis can be maintained throughout life despite age-related changes and the high cyclic loading the tissue sustains. Nevertheless, a large number of factors can cause joint diseases and associated progressive articular cartilage degeneration. These include abnormal loading of a joint due to altered biomechanics (e.g., due to trauma or obesity), as well as degeneration due to genetic, environmental and dietetic factors, among other causes, which may also arise under normal loading conditions [5, 6]. The most common form of joint disease is osteoarthritis (OA), which is estimated to affect approximately 15% of the Canadian population 18 years of age or older [7]. While OA is often characterized in terms of its degenerative effects on articular cartilage, its pathophysiology also involves the other tissues of the affected joint, including the subchondral bone and the synovium [8]. The interplay between these tissues is important in the development and progression of the disease, such that an effective treatment strategy would need to consider and target changes in the joint as a system rather than articular cartilage alone [9, 10]. For example, inflammation of the synovial membrane has been linked with the release of cytokines, such as interleukin 1β (IL-1β) and tumor necrosis factor α, which contribute to the loss of balance between the expression of catabolic enzymes and anabolism in chondrocytes and consequent tissue loss [11].

William Hunter observed more than 250 years ago "when destroyed, [cartilage] is never recovered" [12]. Despite substantial efforts to overcome the limited ability of this tissue to regenerate itself, this early observation holds true today, as there is still no cure available in clinics to interrupt or reverse the progression of OA. As such, treatment modalities aim to either manage the symptoms of the disease or to repair or replace the damaged tissue. Tissue repair is distinct from regeneration: regeneration leads to neo-tissue with similar composition and organization to the native articular cartilage, whereas repair creates tissue with different, typically inferior, properties. Management of symptoms that include swelling, stiffness of the joints and pain is achieved via lifestyle changes through physical exercise and weight loss programs, often combined with the administration of acetaminophen and non-steroidal anti-inflammatory drugs (NSAIDs) [13]. Another non-surgical intervention used to manage the symptoms of cartilage damage is viscosupplementation, the injection of a hyaluronic acid solution into the joint capsule to ease pain and facilitate movement [14]. While disease-modifying OA drugs (DMOAD) remain an unmet need in clinical settings, a number of pharmacological agents targeting OA progression have undergone phase II/III clinical trials in recent years. These include the inducible nitric oxide synthase inhibitor Cindunistat, the IL-1β inhibitor diacerein, oral salmon calcitonin, and strontium ranelate [15–18]. The clinical data from viscosupplementation intra-articular treatments using hyaluronan is also being analyzed to demonstrate chondroprotective effects [19].

A range of surgical procedures is also available to repair cartilage defects when non-invasive approaches are no longer effective for managing symptoms. The gold standard treatment for patients with advanced OA remains the partial or total replacement of the diseased joint with prosthesis. Total joint arthroplasty is typically very successful for reducing pain and improving quality of life [20]. Nevertheless, failure of orthopaedic implants due to infection, fatigue failure or implant loosening caused by wear debris-induced osteolysis requires technically challenging revision surgeries, rendering this approach less suitable for the treatment of younger patients [20, 21]. These limitations of orthopaedic implants have fueled the development of tissue and cell-based interventions. Mosaicplasty, a surgical procedure that consists of harvesting small cylindrical osteochondral samples from low weight-bearing locations of the joint (autologous) or a deceased donor (allogenic) and transplanting these into the defect has shown encouraging clinical results; however, issues related to donor site morbidity, poor integration of the grafts to surrounding tissues, disease transmission and limited treatable defect size have restrained its use [22–24]. The most common surgical approaches are marrow stimulation techniques (e.g., drilling, abrasion and microfracture) [25]. These comprise methodologies to access the bone marrow by breaching the integrity of the subchondral bone and form a fibrin clot within a debrided cartilage defect. Blood-borne progenitor cells can subsequently infiltrate this fibrin "scaffold" and deposit repair tissue. While these involve a single simple intervention, the resulting fibrocartilage repair tissue is typically mechanically inferior to native articular cartilage and is prone to deterioration within a few years [26]. Another approach is autologous chondrocyte implantation (ACI), which involves the excision of autologous cartilage from non-loading areas of the affected joint in a first surgical intervention, followed by the enzymatic release of chondrocytes from their ECM and their amplification in vitro [27]. The chondrocytes are then implanted back into the debrided cartilage defect under a periosteal graft in a second surgery. The long-term clinical outcome for ACI has been positive [28, 29]. A variety of 3D scaffolds and hydrogels have since been developed as cell delivery constructs and/or templates that instruct tissue formation to improve on the results of microfracture and ACI procedures, some of which have been translated into clinical use [30]. The field of cartilage tissue engineering is rapidly evolving and the technologies being developed are increasingly representative of the complexity of the native tissue they aim to regenerate.

Efforts to identify DMOADs and to develop improved scaffolds for cartilage tissue engineering have paralleled increased incorporation of nanomaterials into cartilage repair strategies. According to the definition provided by the ISO/TS 80004-1:2015, nanomaterials are characterized as having at least one of their external dimensions ranging between 1 and 100 nm, or having internal or surface structures in the same range; however, the term is often applied more loosely to sub-micron scale materials in the literature of many fields including tissue engineering (see Chap. 1). At nanoscale (1–100 nm), materials exhibit unique size-dependent properties that arise due to a high surface area-to-volume ratio, resulting in a greater relative contribution from surface molecules compared to those in the bulk. These properties provide interesting opportunities to devise novel strategies for biomedical applications, ranging from biomarker detection and in situ imaging to therapeutic delivery and tissue regeneration. Our understanding of the importance of the nanoscale organization level in tissues also offers inspiration for the development of bioinspired nanomaterials that could deliver improved therapeutic efficacy.

In this chapter, we will briefly review the organization of articular cartilage and its main building blocks from the macroscale down to the nanoscale and discuss submicron particles that participate in cell-cell communication. Efforts to design scaffolds incorporating cell-instructive nanoscale features and to exploit unique nanoparticle properties to design scaffolds and hydrogels with unique properties for cartilage tissue engineering applications will be discussed at some length, e.g, approaches to tailor mechanical properties and to engineer spatial organization in biomaterials with nanomaterials. Finally, key findings for the design of nanoparticles to enable targeted therapeutic delivery will be examined.

5.2   Articular Cartilage from Macro- to Nanoscale

Despite having been portrayed as a rather simple tissue from the point of view of its organization due to the absence of vasculature, lymphatic vessels, and nerves, articular cartilage has multiscale structural complexity. Macroscopically, it appears as a smooth, whitish layer of tissue on the articulating surfaces of long bones, where it reaches thicknesses ranging from less than 1–6 mm depending on anatomical position, age, and exercise level [31, 32].

Articular cartilage exhibits depth-dependent anisotropy as pertains to ECM composition, biomacromolecule organization and cross-linking, as well as resulting mechanical properties [33]. This anisotropy is often discussed in the literature as a "zonal" organization, whereby articular cartilage is divided into four zones; from superior to inferior of the joint surface: the superficial (or tangential) zone, the middle (or transitional) zone, the deep (or radial) zone and the zone of calcified cartilage that interfaces with the subchondral bone [3]. Of note, the collagen fibers are aligned parallel to the tissue surface in the superficial zone, transition to a more random orientation in the middle zone, and are perpendicular to the tissue surface in the deep zone [34] (Fig. 5.1a). A number of studies have proposed the existence of additional distinct structural regions in the most superficial zone of articular cartilage, suggesting a higher level of organizational complexity than is generally appreciated [35, 36]. Owing to the avascular nature of articular cartilage, chondrocytes are supplied with nutrients and signaling molecules by diffusion from the synovial fluid, establishing biomolecular gradients that may contribute to generating and maintaining this anisotropic organization [37, 38]. Zonal differences in the phenotypic specification of chondrocytes also exist, with superficial, middle and deep zone chondrocytes having been characterized [39]. These phenotypic differences result in distinct expression

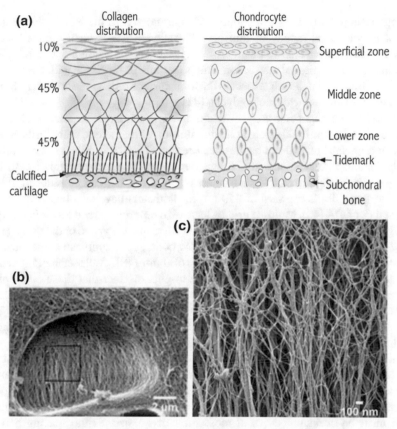

Fig. 5.1 **a** Schematic representation of the zonal organization in articular cartilage. **b, c** Helium ion microscope images of the fibrillar extracellular matrix in the pericellular matrix directly interfaced with chondrocytes. Adapted from Vander Berg-Foels et al. [40] Copyright (2012), with permission from John Wiley and Sons

profiles that dictate the compositional and organizational differences between zones and contribute to the mechanical and tribological functions of articular cartilage.

The articular cartilage ECM is also organized differently with respect to its distance from chondrocytes. The pericellular matrix (PCM) is the thin layer of ECM that directly interfaces each chondrocyte and provides a niche microenvironment for the cells [41]. It has a distinct composition from the remaining cartilage ECM that comprises high proteoglycan content (e.g., perlecan, aggrecan, hyaluronan, and biglycan) and a network of type VI collagen fibrils, a protein that is concentrated in the PCM in cartilage [42]. The chondron, which consists of the chondrocyte and its PCM, has been recognized as the primary biomechanical unit of articular cartilage [43]. Studies suggest that it plays a role in protecting the cell against mechanical loading, as well as in matrix turnover and homeostasis. Important changes have been observed in the mechanical properties and composition of the PCM during OA pathogenesis [44]. Distal from chondrocytes compared to the PCM, the territorial matrix is composed mainly of thin type II collagen fibrils also arranged around chondrocytes and interspersed with proteoglycan to resist loading and deformation, whereas the interterritorial matrix constitutes the bulk of the tissue with larger type II collagen fibrils oriented with respect to the joint surface rather than around individual cells and interspersed with proteoglycan [41].

These ECM components are self-assembled into nanoscale structures. For example, the collagen fibrils exhibit a broad distribution of diameters, with fibrils as small as 10 and upward of 200 nm having been reported [40, 45]. Fibril dimensions are highly dependent on anatomical site, age, depth from the surface, and disease state. For example, their diameter tends to increase with age but can be reduced in early OA, and large fibrils (upward of 450 nm in diameter) have also been observed in later stages of OA [46]. Multiple groups have also reported a trend towards increased collagen fibril diameter from the superficial zone to the deep zone of the tissue [1, 40, 47]. The interfibrillar spaces in cartilage ECM contain proteoglycan aggregates with hydrodynamic radii ranging from 1000 to 1600 nm [48]. A closer look at cellular interactions with the surrounding PCM highlights the fact that these occur at the nanoscale. Indeed, the PCM collagen network in direct contact with chondrocytes is composed of a majority of fibrils with diameters below 100 nm (Fig. 5.1b, c) [40]. This dense arrangement of ECM components represents a considerable resistance to the diffusion of macromolecules within articular cartilage.

The nature of the cartilage ECM combined with its avascular nature and the fact that chondrocytes are relatively isolated and sparse within the tissue, representing only approximately 2% of tissue volume in adults, have implications for cell-to-cell communication mechanisms. Any signaling molecules carrier passing through the ECM must be of nanometer or at most sub-micron size, as the spacing between collagen fibrils in articular cartilage has been reported to range from 60 to 200 nm [45], while the packing of polyanionic glycosaminoglycan subunits in proteoglycan aggregates is even denser with only a few nanometers between branches [48]. The presence of cellular projections connecting chondrocytes within the ECM has been reported, suggesting a potential pathway for direct cell-to-cell communication [49]. An additional proposed means of communication identified in cartilage is the release of extracellular vesicles (EVs) to shuttle cargos of signaling molecules between cells within a tissue and from one tissue to another. EVs consist of three classes of cell-derived particles: exosomes produced by multivesicular endosomes and ranging in size between 30 and 150 nm, microvesicles formed by cell membrane budding with typically larger sizes ranging between 50 and 1000 nm [50] and apoptotic bodies that have been associated with OA [51]. EVs found in cartilage have been studied extensively in the context of ECM calcification [52]. They have also been associated with inter-tissue signaling within the joint [53]. Given the ability of these nanoparticles to diffuse through the dense ECM of articular cartilage and act as delivery vehicles for bioactive molecules to chondrocytes, continued efforts to understand the mechanisms by which EVs are transported through cartilage, including the importance of size and surface properties, may offer a roadmap for achieving delivery and retention of therapeutic cargo within the tissue.

5.3   Nanomaterials in Cartilage Tissue Engineering Scaffolds

One of the fundamental tenets of the tissue engineering approach, when it was first proposed in the 1980s, was the idea that a micro- or macro-porous biocompatible and resorbable material could be used as a scaffold to guide tissue regeneration [54]. A growing appreciation of the importance of carefully tailoring the microenvironment in a scaffold (or template material [55]) to appropriately modulate the phenotype and fate decisions of cells has driven continued innovation in scaffold/template fabrication. Nanotechnology has taken a position at the leading edge of these efforts to design biomimetic and modulatory biomaterials. Nanomaterial fabrication techniques have been exploited to generate structures analogous to ECM at the nanoscale, a critical dimension in cellular sensing. Nanomaterials have enabled tailoring of the mechanical environment via reinforcement, while high conductivity nanoparticles have been used to facilitate the electrical stimulation of cells. These nanoparticles have also been exploited for controlled delivery of biomolecular signals. Advances in each of these categories for applications in cartilage tissue engineering will be discussed in this section.

5.3.1   Nanoscale Structures in Scaffolds

5.3.1.1   Electrospun Scaffolds

Textile fabrication techniques have been explored extensively for the production of fiber-based scaffolds for tissue engineering, affording the opportunity to mimic some aspects of the fibrous nature of the ECM in tissues. One such technique, termed electrospinning, relies on the generation of an electric field between a polymer solution (typically delivered through a needle) and a collector to draw the solution into a fiber. This drawn solution solidifies on its way to the collector as the solvent evaporates to form a membrane of nonwoven material [56]. Through the careful optimization of process parameters that include polymer concentration, solvent selection, polymer solution flow rate, humidity, voltage differential, needle dimensions, as well as the distance between the needle and the collector, one can produce membranes with average fiber diameters ranging from a few micrometers down to the low nanometer range and fairly narrow distributions. Studies have demonstrated the benefits of these materials for cartilage tissue engineering. For example, chondrocytes seeded onto nanofibrous poly(lactic-co-glycolic acid) (PLGA) membranes with an average fiber diameter of 550 nm exhibited increased proliferation and ECM accumulation compared to those cultured onto flat membranes of the same material [57]. Others showed that mesenchymal stem cells (MSC) seeded onto nanofibrous materials made of poly(caprolactone) (PCL) with an average fiber diameter of 700 nm exhibited enhanced chondrogenesis compared to the cell pellet culture model [58].

A number of groups have exploited the control afforded by electrospinning over fibrous material structures to study the effect of fiber diameter on chondrogenic cell responses, with mixed results. Li et al., reported on the interaction of passaged bovine chondrocytes with electrospun poly(L-lactic acid) (PLLA) scaffolds made of nanoscale (500–900 nm) and microscale (15–20 μm) fibers [59]. They showed increased proliferation and sulfated glycosaminoglycan (sGAG) accumulation, as well as decreased dedifferentiation, on the material made of nanofibers, concomitant with more spherical cell morphology. Others who compared PCL membranes with aligned fibers characterized by average dimensions of 500 and 3000 nm also demonstrated a benefit of sub-micron fibers to the gene expression profile of differentiating MSC [60]; however, the opposite trend was also reported in another study that used nonwoven materials made of 440 and 4300 nm fibers [61]. A detailed study comparing a range of different fiber diameters from 300 nm to 9 μm reported that the larger fiber diameter materials elicited increased chondrogenic differentiation in MSC [62]. The authors suggested that this effect might have to do with the increased pore dimensions in scaffolds characterized by micrometer-scale fibers rather than with the smaller fibers. In support of the suggestion that pore size may play an important role in the cellular responses observed between nanofibrous and microfibrous scaffolds, another group found that the generation of multiphasic scaffolds composed of both nanoscale and microscale fibers held benefits (as measured by increased ECM accumulation) compared with scaffolds that only incorporate microfibers [63]. Taken together, these studies highlight the need for additional work to clarify the effects of electrospun fiber dimensions over the full range of sub-micrometer fibers that can be fabricated and the contribution of pore size on the modulation of cell phenotype. Of particular interest is the study of fiber dimensions comparable to those of the fibrous network in articular cartilage. Controlling the diameter of fibers in electrospun scaffolds has also been shown to provide the opportunity to tune mechanical properties of single fibers [64]. This study, based on atomic force microscopy measurements, highlights the fact that the careful selection of the material-dimension combination is critical for presenting cells with a microenvironment that incorporates native biomechanical cues.

The effect of fiber orientation on the phenotype of chondrogenic cells has also been investigated. Aligned electrospun fibers are typically obtained by using a rotating cylindrical mandrel as a collector and adjusting its rotational speed, whereby a faster rotation will result in increased fiber alignment [65]. One study investigated the effect of aligned or randomly oriented sub-micron fibers on the chondrogenic differentiation of human nasal septum-derived progenitor cells cultured on PLLA/PCL blend electrospun membranes [66]. The authors observed that chondrogenic differentiation was enhanced on aligned fibrous materials, whereas cells on randomly oriented membranes showed increased proliferation. Others have explored the use of aligned nanofibers as an approach to specifically engineer the superficial zone of articular cartilage, demonstrating that aligned fibrous structures drive specification into distinct cellular phenotypes compared to other scaffold structures [60, 67, 68]. Other electrospinning techniques have also been developed to achieve nanofiber alignment. For example, a collector consisting of two conducting supports separated by a gap was shown to result in fibers bridging the gap in a highly aligned manner [69]. Others have modified the instrumental setup to stabilize the polymer solution fiber emerging from the Taylor cone at the tip of the dispenser [70]. This setup, combined with a digitally controlled moving collector, allowed the production of spatially-defined electrospun mats exhibiting fiber alignment. Another group has proposed a modified rotating collector presenting a circular surface, which was used to generate electrospun materials with circumferential fiber orientation [71]. The resulting material structure mimicked aspects of the meniscus cartilaginous tissue organization and encouraged alignment of MSC along the changing orientation of the nanofibers.

One important setback in the development of electrospun scaffolds for cartilage tissue engineering applications is the fact that the structures generated are essentially organized on a 2D plane, while the fibrous components of articular cartilage ECM exhibit a complex 3D organization with depth-dependent anisotropy. It should also be emphasized that the relatively small pore size characterizing these scaffolds impedes cell migration through the 3D structure, impacting their potential for applications in the repair/regeneration of full-thickness articular cartilage. This is a particularly important problem for materials made with nanofibers with proportionally smaller pores. This situation has led to a number of innovations in the fabrication procedure and cell seeding protocols. As an example, a modified electrospinning setup was used to deposit nanofibers onto the surface of a microfiber, which was subsequently pressed into the desired scaffold shape and density using a piston [72]. Others have developed an approach that involves co-electrospinning the nanofibrous material and sacrificial fibers that can be dissolved to open up the porous structure [73]. Yet another strategy employed the electrospinning apparatus under conditions that enabled "direct writing" to produce a scaffold consisting of struts oriented in such a way as to recreate the general depth-dependent collagen fiber directionality of native articular cartilage tissue, and subsequently electrospinning a fibrous network onto this open scaffold structure [74]. This approach produces full depth anisotropic articular cartilage scaffolds; however, these represent relatively thin (~200 μm) slices of tissue. Stacking and bonding of multiple slices allowed the generation of large constructs. While micrometer fibers were produced in this study, the approach would be amenable to the application of nanofibers onto the scaffold produced by direct writing.

5.3.1.2 Scaffolds Produced by Phase Separation

Other scaffold fabrication techniques have also been used to produce nanofibrous scaffolds for cartilage tissue engineering, albeit not as extensively as electrospinning. Phase separation is one such fabrication technique that exploits the fact that a homogenous polymer solution will separate into polymer-rich and polymer-poor phases in thermodynamically unstable conditions, including during temperature-induced solidification. Once the solution is frozen, the resulting material can be sublimed to remove the solvent and reveal a porous structure in place of the polymer-poor regions, whereby polymer-rich regions form the scaffold walls. This methodology has been modified to include a gelation step prior to freezing, which results in nanofibrous structures with fiber diameters ranging from 50 to 500 nm [75]. Furthermore, this method offers the opportunity to produce fiber networks in 3D, in contrast with the planar arrangement of electrospun membranes. This fabrication technique has been applied to produce scaffolds for cartilage tissue engineering. For example, Ma and colleagues combined thermally-induced phase separation (TIPS) with solvent casting porogen leaching to generate scaffolds with controllable micron-scale pores and nanofibrous walls [76, 77] (Fig. 5.2). In these studies, sacrificial spherical sugar particles were sintered prior to solvent casting, gelation and freeze-drying to form interconnected pores. A similar approach was used to compare the effect of nanofibrous versus dense scaffold walls on chondrocytes, demonstrating more rounded cell morphology, improved chondrogenic phenotype and increased ECM accumulation on the nanofibrous scaffolds [78]. Another study combined TIPS with 3D printing to yield scaffolds with both macroscopic architecture and nanoscale features [79]. This multiscale scaffold resulted in substantially increased cell adhesion and accumulation of key ECM components (sGAG and collagen). Another group proposed a different strategy to generate nanofibrous scaffolds, employing TIPS with mixture of PLLA and camphene that forms an interpenetrating network [80]. Because of its physical properties, camphene can be removed during the sublimation step to reveal the nanofibrous structure. In this study, chondrocytic phenotype was reduced for

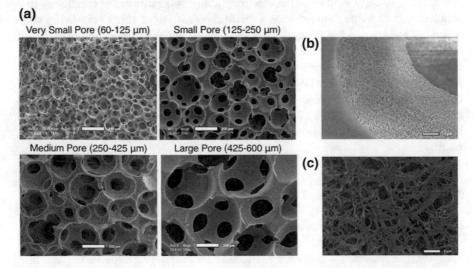

Fig. 5.2 Scanning electron micrographs of scaffolds fabricated by thermally-induced phase separation combined with solvent casting porogen leaching to generate nanofibrous scaffolds walls. Reprinted from Gupte et al. [77] Copyright (2018), with permission from Elsevier

cells in the nanofibrous scaffold compared to cells cultured in scaffolds produced in absence of camphene and characterized by dense, non-fibrous walls. The authors proposed that these results were due to increased cell-material interaction in the nanofibrous scaffold.

5.3.1.3 Self-assembled Supramolecular Structures

The spontaneous self-assembly of molecular building blocks into nanoscale structures, including nanofibers, has also been exploited as a promising strategy for biomaterials fabrication [81]. Through the careful design of these molecular building blocks, often inspired by self-assembly processes occurring in nature, control over intermolecular interactions can be achieved such that supramolecular architectures can be generated. A range of biomolecules that includes peptides, DNA and lipids has been self-assembled and stabilized through non-covalent forces including hydrogen bonds, ionic interactions, and Van der Waals forces. Typically, the fibers produced by self-assembly have diameters of 10 nm or less, representing the lower end of those found in articular cartilage ECM [81, 82]. This also represents a length scale that is more difficult to achieve with electrospinning and phase separation.

The concept of self-assembly has found applications in the design of novel scaffold and hydrogel structures for cartilage tissue engineering applications. Stupp and colleagues have produced self-assembled nanofiber gels from tailored peptide amphiphiles functionalized with transforming growth factor β-1 (TGFβ-1) binding peptide [83]. This gel allowed for a slower release of the growth factor than was observed with non-functionalized materials. The constructs were tested in full-thickness chondral defects treated with microfracture in rabbits to encourage bone marrow stromal cells into the defect. The authors observed that the functionalized gels enhanced articular cartilage regeneration compared to control groups that received an injection of TGFβ-1 or a non-functionalized gel loaded with the growth factor. Others have combined decellularized cartilage matrices (DCM) with nanofibrous gels of self-assembled peptide. Improved cartilage regeneration was demonstrated following implantation of the combined gel/DCM scaffold in rabbit full-thickness defects treated with microfracture compared to microfracture alone, as well as when compared with the microfracture plus the decellularized cartilage matrix scaffold absent the self-assembled nanofibrous gel [84]. Other building blocks have also been used to create constructs incorporating self-assembly. For example, self-assembled DNA-based rosette nanotubes functionalized with the integrin-binding peptide motif RGDSK have been used to produce nanofibrous scaffolds. These scaffolds were shown to support the chondrogenic differentiation of human MSC [85]. These studies highlight the potential of self-assembled nanofibrous scaffolds and hydrogels for cartilage tissue engineering. These highly tunable materials offer the opportunity to tailor the complexity of the microenvironments presented to resident cells and further instruct their responses.

5.3.1.4   Scaffold Surface Nanoroughness

Recognizing the importance of biomimetic surface roughness for controlling cell responses, many groups have proposed surface modification treatments to incorporate topographical features on biomaterials. These techniques have been explored in greater detail for metallic orthopedic implants interfaced with bone; nevertheless, a few studies have investigated the effects of nanoroughness on chondrocyte response. For example, Webster and colleagues used a short NaOH immersion treatment to modify the surface roughness of PLGA scaffolds produced by porogen leaching [86]. The resulting surface topography led to increased chondrocyte attachment, growth and ECM accumulation compared to untreated surfaces. The same group also developed a method to generate nanoroughness on the surface of polyurethane and PCL films by casting the polymer solution onto plasma modified titanium [87]. This surface modification also led to increased chondrocyte attachment and intracellular collagen content. Both of these surface modification protocols caused concurrent production of micro- and nanoscale surface modifications; the extent to which nanoroughness contributed to the observed effects, therefore, remains unclear. In a more fundamental study on the topic, nanotopographical features (nano-pillar, nano-hole, and nano-groove arrays) were produced on PCL surfaces by thermal nanoimprinting [88]. The effects of these surfaces on MSC response were investigated against non-modified surfaces. Nano-pillar and nano-hole arrays exhibited decreased cell proliferation and increased chondrogenic differentiation compared to cells cultured on control surfaces, while nano-grooves led to cell elongation and encouraged phenotypic changes reminiscent of superficial zone chondrocytes. These studies highlight the need for additional work in this area, in order to assess in greater detail how nanoscale topography and roughness can be incorporated with cartilage repair scaffolds to direct cellular responses, as well as responses to microscale versus nanoscale surface modifications.

5.4   Nanoparticle Composites to Tailor Mechanical Microenvironment

5.4.1   Tuning Mechanical Properties

As described previously, articular cartilage is a load-bearing tissue that fulfills primarily biomechanical functions in the body. As such, it displays relatively high mechanical properties compared to other soft tissues and behaves as a viscoelastic material with creep and stress relaxation responses [89]. Efforts to design scaffolds that mimic key aspects of native tissue's mechanical properties have typically involved trade-offs with scaffold porosity. In the same way, biocompatible hydrogel systems often exhibit mechanical properties that are orders of magnitude lower than those of the native tissue. To address this limitation, nanoparticles have been added to scaffolds and hydrogels to generate mechanically reinforced composites. A number of different nanoparticles have been used for this purpose in cartilage tissue engineering thus far, including carbon nanotubes [90], Laponite clay particles [91], poly(styrene-acrylic acid) core-shell particles [92], and cellulose nanocrystals [93], to name a few. The reinforcement effect of nanoparticles has been associated with their ability to interact with polymer molecules and form additional cross-links within the resulting structure. An increased surface area leads to greater interaction with the surrounding hydrogel, and nanoscale particles offer the advantage that they exhibit a substantially increased surface area to weight ratio compared to larger particles [94].

5.4.2 Enabling Mechanical Stimulation

The average person takes approximately 2 million steps per year; that is to say that the joints in our leg each typically undergo 1 million loading cycles annually [95]. The importance of mechanical loading in articular cartilage remodeling is well established. Indeed, vigorous physical activity in healthy individuals has been associated with increased cartilage volume and a decreased risk of developing cartilage defects [96]. Biomechanical factors including obesity and injuries leading to joint instabilities, on the other hand, have been associated with increased risks of cartilage pathologies [97]. In vitro studies on chondrocyte and cartilage response to mechanical loading have revealed ranges of stimulation parameters that result in increased tissue formation, while deviation from appropriate loading frequency, strain rate, and amplitude, as well as the loading history have been associated with increased catabolic responses [98]. Substantial efforts in the field have therefore focused on the development of bioreactors to facilitate the application of biologically relevant biomechanical stimulation regimens to induce increased tissue formation in engineered constructs.

Magnetic nanoparticles have been used to stimulate constructs mechanically. These can be incorporated into cells or materials and exposed to a magnetic field to induce strain. For example, magnetic nanoparticles synthesized by *Magnetospirillum sp.* AMB-1 can be efficiently endocytosed by MSC. Exposing the treated to MSC to magnetic fields leads to the application of forces to the cells [99]. Here, the authors showed significantly increased ECM (sGAG and collagen) accumulation and chondrogenic gene expression, in cell pellets subjected to short term physical stimulation (1 h per day for 5 consecutive days) compared to controls at 3 weeks post-stimulation. Although the nanoparticles were not found to be cytotoxic at concentrations below 30 μg/ml, three times above the levels required to achieve cellular magnetization, the long-term safety and clearance of these nanoparticles remains to be clarified. Other groups have incorporated magnetic nanoparticles into hydrogel materials [100, 101]. Ethier and colleagues produced trilayered hydrogels with each zone characterized by a specific agarose concentration and nanoparticle loading [101]. With this approach, the authors were able to produce differential strains in each zone of the construct, mimicking the anisotropic response of native articular cartilage to mechanical loading. In an interesting study, magnetic nanoparticles were functionalized with an antibody against Frizzled, a receptor for the Wnt signaling pathway, which is of importance in chondrogenesis [102]. These functionalized nanoparticles were incubated with human MSC and shown to bind the Frizzled receptors on their surface. An oscillating magnetic bioreactor was then used to mechanically stimulate the receptor and activate the Wnt pathway. Such an approach could prove powerful for cartilage tissue engineering applications, whereby specific mechanosensitive signaling pathways may be activated without relying on chemicals or drugs. A similar approach had previously been used to activate the potassium channel TREK-1 on the surface of human MSC both in vitro and in vivo, resulting in the upregulation of genes associated with both osteogenesis and chondrogenesis, as well as increased synthesis of ECM components [103]. It should also be mentioned that the application of magnetic fields to chondrocyte cultures, even in the absence of magnetic nanoparticles, can cause cellular responses such as increased proliferation and increased sGAG accumulation [104, 105].

Park and colleagues have proposed an alternative application of magnetic nanoparticles for cartilage tissue engineering [106]. In this study, the authors produced porous microbead-shaped PLGA scaffolds. They used water-in-oil-in-water emulsion templating to achieve microbead structures containing gelatin particles and subsequently leached the gelatin to achieve porosity (Fig. 5.3). The surfaces of the resulting microscaffolds were further functionalized with Fe_3O_4 magnetic nanoparticles, enabling their actuation and deployment to a site of injury under the influence of a magnetic field. These scaffolds were shown to support attachment and chondrogenic differentiation of MSC. Such an approach has the potential for minimally-invasive surgical treatment of joint ailments.

5.4.3   Enabling Electrical Stimulation

Chondrocytes are not considered excitable cells in the same way that neurons and myocytes are; however, these cells are particular in that they exist in a higher osmolarity microenvironment than many other cell types [107]. Furthermore, increasing evidence points to the importance of calcium, sodium, and potassium signaling in chondrocyte and cartilage homeostasis through a complex channelome [108]. Given the importance of charged metal cations in chondrocyte signaling and the presence of voltage-gated ion channels on chondrocyte membrane, electrical stimulation has been investigated extensively for the treatment of cartilage ailments [109]. The responsiveness of chondrocytes to electrical signals was exploited by Webster and colleagues, who demonstrated that loading of conductive carbon nanotubes into polyurethane and subsequent electrical stimulation through the polymer enhanced both chondrocyte adhesion and proliferation compared to neat (unloaded) polymer [110]. The authors further demonstrated that the effect was not only caused by the electrical stimulation and was also due in part to the increased surface nanoroughness resulting from the incorporation of the nanoparticles within the polymer sheets.

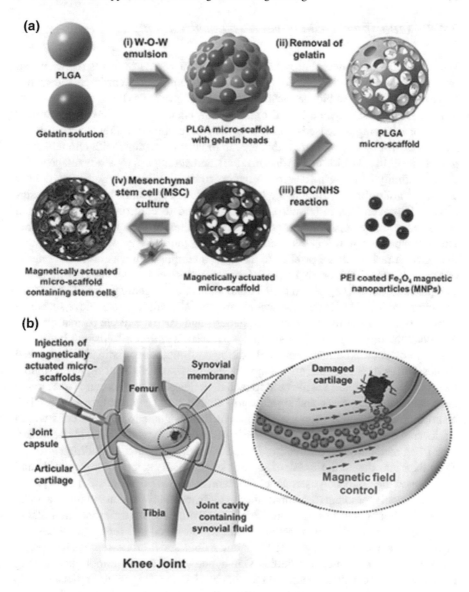

Fig. 5.3 Schematic diagram of a micro-scaffold fabrication process that incorporates magnetic nanoparticles to enable actuation and deployment into cartilage defects. Reprinted from Go et al. [106] Copyright (2017), with permission from John Wiley and Sons

5.4.4   Tailoring the Chemical Microenvironment

While tailoring the physical properties of cellular microenvironments represents an important design consideration for the formation of engineered cartilage, the spatio-temporal availability of biochemical signals has also proven to be equally important in stimulating tissue synthesis and organization. Nanomaterials have been instrumental in achieving increased control over the presentation of biomolecular signals, such as growth factors and small therapeutic molecules, to cells with chondrogenic potential within scaffolds and hydrogels. Park and colleagues took advantage of the specific affinity of heparin for growth factors to develop nanoparticles that deliver TGF-β3 within fibrin hydrogels seeded with MSC [111]. This construct led to significant improvements in chondrogenesis compared to controls, as well as when compared with hydrogels incorporating nanoparticles or the growth factor alone. Other groups have developed systems comprising multiple nanoparticles with distinct growth factor release profiles to integrate a temporal dimension to the release of a suite of growth factors. Nanoparticles have also served to deliver growth factor-rich platelet lysate [112], plasmid DNA-encoding chondrogenic growth factor [113] and bioactive ions [114]. These strategies are typically tailored to achieve sustained delivery of important factors in chondrogenesis and cartilage tissue formation. Furthermore, the uniform distribution of nanoparticles within scaffolds and hydrogels allows to overcome biomacromolecule diffusion limitations within 3D engineered tissues. Diffusion limitations are a major problem in tissue engineering when soluble factors are administered via the culture media, as these limitations can lead to tissue deposition inhomogeneity and significantly altered cellular phenotypes [115, 116]. Nanoparticles have also been used to present ECM signals to resident cells within engineered constructs. This is illustrated by Gibson et al, who produced decellularized ECM nanoparticles originating from a number of tissues, including cartilage. They then introduced the ECM nanoparticles into PCL electrospun scaffolds and investigated the effects on osteogenesis of human adipose-derived stem cells [117].

Biomolecular gradients are important signaling mechanisms that modulate a broad range of cellular responses from proliferation and migration to differentiation. These gradients play crucial roles in development, maintenance and repair of tissues and organs, while also being implicated in many pathological processes. As was previously discussed, biomolecular gradients are hypothesized to play important signaling roles in articular cartilage, owing to fact that nutrients and signaling molecules gain access to the tissue primarily via its superficial aspect (i.e, its surface; see Fig. 5.1a). Generating biologically relevant biomolecular gradients within scaffolds and hydrogels to direct anisotropic tissue organization represents a long-standing challenge in tissue engineering. This is an area where nanoparticles have had an important impact. For example, hydroxyapatite nanoparticles stimulate the osteogenic differentiation of MSC [118]. A number of groups have exploited this effect to generate MSC-containing scaffolds and hydrogels with spatially constrained nanoscale hydroxyapatite particles, and thus drive osteogenesis locally, while encouraging chondrogenesis in areas devoid of nanoparticles [119, 120]. In this way, biphasic constructs

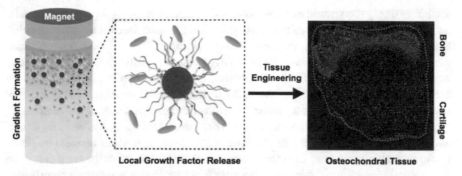

Fig. 5.4 Schematic diagram of procedure to generate biochemical gradients across the depth of hydrogels with heparin-functionalized, growth factor-loaded superparamagnetic nanoparticles. Reprinted from Li et al. [122] Copyright (2018), according to a Creative Commons Attribution License (CC BY)

containing both cartilage and bone, reminiscent of the osteochondral organization present in joints, can be achieved. Radhakrishnan et al. recently proposed a biphasic construct generated by spatially localizing hydroxyapatite and chondroitin sulfate nanoparticles within an alginate poly(vinyl alcohol) hydrogel. The zone loaded with hydroxyapatite nanoparticles generated subchondral bone tissue, while the chondroitin sulfate particles induced cartilage) tissue formation, leading to production of an integrated osteochondral construct [121]. Stevens and colleagues proposed various approaches to generate biomolecular gradients that can be exploited to produce osteochondral hydrogel constructs. In a first study, superparamagnetic nanoparticles were surface-functionalized with heparin, which acted as a reservoir for bone morphogenetic protein 2 [122]. These loaded nanoparticles were incorporated with hydrogel precursors, and the resulting solution was subjected to a magnetic field during the hydrogel cross-linking step. The process generated biomolecular gradients within the hydrogel (Fig. 5.4). In a second study, buoyancy was used to drive the formation of gradients with different types of nanoparticles in a range of hydrogel base materials [123].

5.5 Nanoparticles for Drug Delivery

Many patients with cartilage and joint ailments exhibit advanced signs of articular cartilage degeneration, or substantial injuries to their articular surface that are deemed to be at high risk of degeneration. The recommended course of action is in these cases is typically a surgical intervention. Other patients present early signs of degeneration for which a more conservative approach is favored. For these patients, a number of non-pharmaceutical and pharmaceutical options are available; however, as detailed previously, these therapeutics are aimed at managing symptoms and DMOAD are still an unmet need. Issues with the bioavailability of drugs within the joint space are increasingly recognized as an important factor in explaining the absence of safe and efficacious DMOAD despite intense efforts in the field. Because of the avascular nature of articular cartilage, the target tissue for many DMOAD candidates, local administration of drugs has been favoured over systemic delivery strategies. Indeed, intra-articular injection provides the opportunity to bypass barriers to drug transport across vascular walls, as well as the ECM of the synovial membrane, and into the synovial fluid. As such, intra-articular injections have been associated with increased local bioavailability for a given administered drug dosage, reduced systemic exposure and thus decreased off-target effects. However, these require administration by practitioners, making this drug delivery strategy costlier and logistically more complex than self-administration strategies, especially for chronic conditions such as arthritis, which require sustained treatment over a period of years to decades. Furthermore, synovial fluid turnover is rapid and injected molecules are typically removed from the intra-articular space via lymphatic drainage in a manner of hours, such that maintaining drug levels within their therapeutic window in the joint is often impractical in clinical settings. Efforts have consequently centered on the development of strategies to increase the retention time of therapeutics within the synovial capsule following intra-articular injection, notably with injectable hydrogels, microcarriers, and nanoparticles. Nanoparticles offer a unique opportunity for drug delivery in the joint as demonstrated by Hubbell and colleagues, who proposed using the articular cartilage matrix as a reservoir for therapeutic molecules and developed nanoparticles that were small enough to penetrate the small pores of articular cartilage and accumulate in its ECM, as well as intracellularly [124]. The authors functionalized the nanoparticles with a short type II collagen-binding peptide that had been identified via phage display to achieve prolonged retention in the cartilage. Since this early effort, a broad range of nanocarriers have been proposed, including cationic and polyelectrolyte nanoparticles, which have exploited the polyanionic nature of the proteoglycan compartments of cartilage to achieve important penetration depths [125].

5.6   Concluding Remarks

Nanomaterials exhibit a host of unique properties due to the increased relative contribution of surface molecules in relation to those composing the bulk material. Furthermore, there is an increased appreciation of the importance of tissue organization at the nanometer scale for cell and tissue functions. These factors have found many applications in tissue engineering and efforts to repair or regenerate articular cartilage are no exception. However, the incorporation of nanomaterials for the regeneration of articular cartilage remains an emerging strategy. A number of techniques that have been thoroughly investigated in other tissue systems have yet to be explored in-depth for articular cartilage. Some of these areas have been highlighted in this chapter. As this field continues to mature, nanomaterial cartilage tissue engineering will undoubtedly help deliver a range of therapeutic solutions to address joint ailments for a broad spectrum of patients and conditions, from improved early interventions to slow the progress of the disease to the development of implantable materials to resurface damaged and diseased joints.

Acknowledgements The authors are thankful to Ms. Allison Simmonds for her numerous revisions and comments on the chapter. JPS acknowledges funding from the Natural Sciences and Engineering Research Council of Canada (NSERC) through the Discovery Grant and financial support by the University of Ottawa Seed Funding Opportunity Grant.

Disclosure All authors have read and approved this final version.

References

1. Buckwalter JA, Mankin HJ. Articular cartilage: tissue design and chondrocyte-matrix interactions. Instr Course Lect. 1998;47:477–86.
2. Sophia Fox AJ, Bedi A, Rodeo SA. The basic science of articular cartilage: structure, composition, and function. Sports Health. 2009;1(6):461–8.
3. Poole AR, Kojima T, Yasuda T, Mwale F, Kobayashi M, Laverty S. Composition and structure of articular cartilage: a template for tissue repair. Clin Orthop Relat Res. 2001;(391 Suppl):S26–33.
4. Akkiraju H, Nohe A. Role of chondrocytes in cartilage formation, progression of osteoarthritis and cartilage regeneration. J Dev Biol. 2015;3(4):177–92.
5. Goldring MB, Goldring SR. Osteoarthritis. J Cell Physiol. 2007;213(3):626–34.
6. Lorenz H, Richter W. Osteoarthritis: cellular and molecular changes in degenerating cartilage. Prog Histochem Cytochem. 2006;40(3):135–63.
7. Plotnikoff R, Karunamuni N, Lytvyak E, Penfold C, Schopflocher D, Imayama I, Johnson ST, Raine K. Osteoarthritis prevalence and modifiable factors: a population study. BMC Public Health. 2015;15:1195.
8. Brandt KD, Radin EL, Dieppe PA, van de Putte L. Yet more evidence that osteoarthritis is not a cartilage disease. Ann Rheum Dis. 2006;65(10):1261–4.
9. Scanzello CR, Goldring SR. The role of synovitis in osteoarthritis pathogenesis. Bone. 2012;51(2):249–57.
10. Sharma AR, Jagga S, Lee SS, Nam JS. Interplay between cartilage and subchondral bone contributing to pathogenesis of osteoarthritis. Int J Mol Sci. 2013;14(10):19805–30.
11. Fernandes JC, Martel-Pelletier J, Pelletier JP. The role of cytokines in osteoarthritis pathophysiology. Biorheology. 2002;39(1–2):237–46.
12. Hunter W. Of the structure and diseases of articu-lating cartilages. Philos Trans. 1743;42(470):514–21.
13. Clouet J, Vinatier C, Merceron C, Pot-vaucel M, Maugars Y, Weiss P, Grimandi G, Guicheux J. From osteoarthritis treatments to future regenerative therapies for cartilage. Drug Discov Today. 2009;14(19–20):913–25.
14. Wang CT, Lin J, Chang CJ, Lin YT, Hou SM. Therapeutic effects of hyaluronic acid on osteoarthritis of the knee. A meta-analysis of randomized controlled trials. J Bone Joint Surg Am. 2004;86-A(3):538–45.
15. Hellio le Graverand MP, Clemmer RS, Redifer P, Brunell RM, Hayes CW, Brandt KD, Abramson SB, Manning PT, Miller CG, Vignon E. A 2-year randomised, double-blind, placebo-controlled, multicentre study of oral selective iNOS inhibitor, cindunistat (SD-6010), in patients with symptomatic osteoarthritis of the knee. Ann Rheum Dis. 2013;72(2):187–95.
16. Karsdal MA, Byrjalsen I, Alexandersen P, Bihlet A, Andersen JR, Riis BJ, Bay-Jensen AC, Christiansen C, investigators CC. Treatment of symptomatic knee osteoarthritis with oral salmon calcitonin: results from two phase 3 trials. Osteoarthr Cartilage. 2015;23(4):532–43.

17. Pavelka K, Bruyere O, Cooper C, Kanis JA, Leeb BF, Maheu E, Martel-Pelletier J, Monfort J, Pelletier JP, Rizzoli R and others. Diacerein: benefits, risks and place in the management of osteoarthritis. An opinion-based report from the ESCEO. Drugs Aging. 2016;33(2):75–85.

18. Reginster JY, Badurski J, Bellamy N, Bensen W, Chapurlat R, Chevalier X, Christiansen C, Genant H, Navarro F, Nasonov E and others. Efficacy and safety of strontium ranelate in the treatment of knee osteoarthritis: results of a double-blind, randomised placebo-controlled trial. Ann Rheum Dis. 2013;72(2):179–86.

19. Yu SP, Hunter DJ. Intra-articular therapies for osteoarthritis. Expert Opin Pharmacother. 2016;17(15):2057–71.

20. Bauer TW, Schils J. The pathology of total joint arthroplasty. I. Mechanisms of implant fixation. Skeletal Radiol. 1999;28(8):423–32.

21. Bauer TW, Schils J. The pathology of total joint arthroplasty. II. Mechanisms of implant failure. Skeletal Radiol. 1999;28(9):483–97.

22. Hangody L, Kish G, Karpati Z, Szerb I, Udvarhelyi I. Arthroscopic autogenous osteochondral mosaicplasty for the treatment of femoral condylar articular defects. A preliminary report. Knee Surg Sports Traumatol Arthrosc. 1997;5(4):262–7.

23. Szerb I, Hangody L, Duska Z, Kaposi NP. Mosaicplasty: long-term follow-up. Bull Hosp Jt Dis. 2005;63(1–2):54–62.

24. Bartha L, Vajda A, Duska Z, Rahmeh H, Hangody L. Autologous osteochondral mosaicplasty grafting. J Orthop Sports Phys Ther. 2006;36(10):739–50.

25. Steadman JR, Rodkey WG, Rodrigo JJ. Microfracture: surgical technique and rehabilitation to treat chondral defects. Clin Orthop Relat Res. 2001;(391 Suppl):S362–9.

26. Shapiro F, Koide S, Glimcher MJ. Cell origin and differentiation in the repair of full-thickness defects of articular cartilage. J Bone Joint Surg Am. 1993;75(4):532–53.

27. Brittberg M, Lindahl A, Nilsson A, Ohlsson C, Isaksson O, Peterson L. Treatment of deep cartilage defects in the knee with autologous chondrocyte transplantation. N Engl J Med. 1994;331(14):889–95.

28. Brittberg M, Peterson L, Sjogren-Jansson E, Tallheden T, Lindahl A. Articular cartilage engineering with autologous chondrocyte transplantation. A review of recent developments. J Bone Joint Surg Am. 2003;85-A(Suppl 3):109–15.

29. Peterson L, Vasiliadis HS, Brittberg M, Lindahl A. Autologous chondrocyte implantation: a long-term follow-up. Am J Sports Med. 2010;38(6):1117–24.

30. Vinatier C, Guicheux J. Cartilage tissue engineering: from biomaterials and stem cells to osteoarthritis treatments. Ann Phys Rehabil Med. 2016;59(3):139–44.

31. Cohen ZA, McCarthy DM, Kwak SD, Legrand P, Fogarasi F, Ciaccio EJ, Ateshian GA. Knee cartilage topography, thickness, and contact areas from MRI: in-vitro calibration and in-vivo measurements. Osteoarth Cartilage. 1999;7(1):95–109.

32. Pollock J, O'Toole RV, Nowicki SD, Eglseder WA. Articular cartilage thickness at the distal radius: a cadaveric study. J Hand Surg Am. 2013;38(8):1477–81. Discussion 1482-3.

33. Bergholt MS, St-Pierre JP, Offeddu GS, Parmar PA, Albro MB, Puetzer JL, Oyen ML, Stevens MM. Raman spectroscopy reveals new insights into the zonal organization of native and tissue-engineered articular cartilage. ACS Cent Sci. 2016;2(12):885–95.

34. Jeffery AK, Blunn GW, Archer CW, Bentley G. Three-dimensional collagen architecture in bovine articular cartilage. J Bone Joint Surg Br. 1991;73(5):795–801.

35. Fujioka R, Aoyama T, Takakuwa T. The layered structure of the articular surface. Osteoarthr Cartilage. 2013;21(8):1092–8.

36. MacConaill MA. The movements of bones and joints; fundamental principles with particular reference to rotation movement. J Bone Joint Surg Br. 1948;30B(2):322–6.

37. Spitters TW, Leijten JC, Deus FD, Costa IB, van Apeldoorn AA, van Blitterswijk CA, Karperien M. A dual flow bioreactor with controlled mechanical stimulation for cartilage tissue engineering. Tissue Eng Part C Methods. 2013;19(10):774–83.

38. Zhou S, Cui Z, Urban JP. Factors influencing the oxygen concentration gradient from the synovial surface of articular cartilage to the cartilage-bone interface: a modeling study. Arthritis Rheum. 2004;50(12):3915–24.

39. Coates EE, Fisher JP. Phenotypic variations in chondrocyte subpopulations and their response to in vitro culture and external stimuli. Ann Biomed Eng. 2010;38(11):3371–88.
40. Vanden Berg-Foels WS, Scipioni L, Huynh C, Wen X. Helium ion microscopy for high-resolution visualization of the articular cartilage collagen network. J Microsc. 2012;246(2):168–76.
41. Buckwalter JA, Mankin HJ, Grodzinsky AJ. Articular cartilage and osteoarthritis. Instr Course Lect. 2005;54:465–80.
42. Wilusz RE, Sanchez-Adams J, Guilak F. The structure and function of the pericellular matrix of articular cartilage. Matrix Biol. 2014;39:25–32.
43. Benninghoff A. Form und Bau der Gelenkknorpel in ihren Beziehungen zur Funktion. Der Aufbau des Gelenkknorpels in seinen Beziehungen zur Funktion. Zeitschriftfar Zellforschung und mikroskopische Anatomie. 1925;2(5):783–862.
44. Guilak F, Nims RJ, Dicks A, Wu CL, Meulenbelt I. Osteoarthritis as a disease of the cartilage pericellular matrix. Matrix Biol. 2018;71–72:40–50.
45. Hall BK, Newman S. Cartilage: molecular aspects. Boca Raton, FL: CRC Press; 1991. p. 268.
46. Weiss C. Ultrastructural characteristics of osteoarthritis. Fed Proc. 1973;32(4):1459–66.
47. Poole AR, Pidoux I, Reiner A, Rosenberg L. An immunoelectron microscope study of the organization of proteoglycan monomer, link protein, and collagen in the matrix of articular cartilage. J Cell Biol. 1982;93(3):921–37.
48. Comper WD, Williams RP. Hydrodynamics of concentrated proteoglycan solutions. J Biol Chem. 1987;262(28):13464–71.
49. Mayan MD, Gago-Fuentes R, Carpintero-Fernandez P, Fernandez-Puente P, Filgueira-Fernandez P, Goyanes N, Valiunas V, Brink PR, Goldberg GS, Blanco FJ. Articular chondro-cyte network mediated by gap junctions: role in metabolic cartilage homeostasis. Ann Rheum Dis. 2015;74(1):275–84.
50. Tkach M, Thery C. Communication by extracellular vesicles: where we are and where we need to go. Cell. 2016;164(6):1226–32.
51. Hashimoto S, Ochs RL, Komiya S, Lotz M. Linkage of chondrocyte apoptosis and cartilage degradation in human osteoarthritis. Arthritis Rheum. 1998;41(9):1632–8.
52. Golub EE. Role of matrix vesicles in biomineralization. Biochim Biophys Acta. 2009;1790(12):1592–8.
53. Miyaki S, Lotz MK. Extracellular vesicles in cartilage homeostasis and osteoarthritis. Curr Opin Rheumatol. 2018;30(1):129–35.
54. Langer R, Vacanti JP. Tissue engineering. Science. 1993;260(5110):920–6.
55. Williams DF. The biomaterials conundrum in tissue engineering. Tissue Eng Part A. 2014;20(7–8):1129–31.
56. Braghirolli DI, Steffens D, Pranke P. Electrospinning for regenerative medicine: a review of the main topics. Drug Discov Today. 2014;19(6):743–53.
57. Shin HJ, Lee CH, Cho IH, Kim YJ, Lee YJ, Kim IA, Park KD, Yui N, Shin JW. Electrospun PLGA nanofiber scaffolds for articular cartilage reconstruction: mechanical stability, degradation and cellular responses under mechanical stimulation in vitro. J Biomater Sci Polym Ed. 2006;17(1–2):103–19.
58. Li WJ, Tuli R, Okafor C, Derfoul A, Danielson KG, Hall DJ, Tuan RS. A three-dimensional nanofibrous scaffold for cartilage tissue engineering using human mesenchymal stem cells. Biomaterials. 2005;26(6):599–609.
59. Li WJ, Jiang YJ, Tuan RS. Chondrocyte phenotype in engineered fibrous matrix is regulated by fiber size. Tissue Eng. 2006;12(7):1775–85.
60. Wise JK, Yarin AL, Megaridis CM, Cho M. Chondrogenic differentiation of human mes-enchymal stem cells on oriented nanofibrous scaffolds: engineering the superficial zone of articular cartilage. Tissue Eng Part A. 2009;15(4):913–21.
61. Bean AC, Tuan RS. Fiber diameter and seeding density influence chondrogenic differentia-tion of mesenchymal stem cells seeded on electrospun poly(epsilon-caprolactone) scaffolds. Biomed Mater. 2015;10(1):015018.

62. Shanmugasundaram S, Chaudhry H, Arinzeh TL. Microscale versus nanoscale scaffold architecture for mesenchymal stem cell chondrogenesis. Tissue Eng Part A. 2011;17(5–6):831–40.

63. Levorson EJ, Raman Sreerekha P, Chennazhi KP, Kasper FK, Nair SV, Mikos AG. Fabrication and characterization of multiscale electrospun scaffolds for cartilage regeneration. Biomed Mater. 2013;8(1):014103.

64. Jankovic B, Pelipenko J, Skarabot M, Musevic I, Kristl J. The design trend in tissue-engineering scaffolds based on nanomechanical properties of individual electrospun nanofibers. Int J Pharm. 2013;455(1–2):338–47.

65. Li WJ, Mauck RL, Cooper JA, Yuan X, Tuan RS. Engineering controllable anisotropy in electrospun biodegradable nanofibrous scaffolds for musculoskeletal tissue engineering. J Biomech. 2007;40(8):1686–93.

66. Shafiee A, Seyedjafari E, Sadat Taherzadeh E, Dinarvand P, Soleimani M, Ai J. Enhanced chondrogenesis of human nasal septum derived progenitors on nanofibrous scaffolds. Mater Sci Eng C Mater Biol Appl. 2014;40:445–54.

67. Owida HA, Yang R, Cen L, Kuiper NJ, Yang Y. Induction of zonal-specific cellular morphology and matrix synthesis for biomimetic cartilage regeneration using hybrid scaffolds. J R Soc Interface. 2018;15(143).

68. Schneider T, Kohl B, Sauter T, Kratz K, Lendlein A, Ertel W, Schulze-Tanzil G. Influence of fiber orientation in electrospun polymer scaffolds on viability, adhesion and differentiation of articular chondrocytes. Clin Hemorheol Microcirc. 2012;52(2–4):325–36.

69. Li D, Wang YL, Xia YN. Electrospinning of polymeric and ceramic nanofibers as uniaxially aligned arrays. Nano Lett. 2003;3(8):1167–71.

70. Kim GH. Electrospun PCL nanofibers with anisotropic mechanical properties as a biomedical scaffold. Biomed Mater. 2008;3(2).

71. Fisher MB, Henning EA, Soegaard N, Esterhai JL, Mauck RL. Organized nanofibrous scaffolds that mimic the macroscopic and microscopic architecture of the knee meniscus. Acta Biomater. 2013;9(1):4496–504.

72. Thorvaldsson A, Stenhamre H, Gatenholm P, Walkenstrom P. Electrospinning of highly porous scaffolds for cartilage regeneration. Biomacromol. 2008;9(3):1044–9.

73. Baker BM, Shah RP, Silverstein AM, Esterhai JL, Burdick JA, Mauck RL. Sacrificial nanofibrous composites provide instruction without impediment and enable functional tissue formation. Proc Natl Acad Sci USA. 2012;109(35):14176–81.

74. Chen HL, Malheiro ADFB, van Blitterswijk C, Mota C, Wieringa PA, Moroni L. Direct writing electrospinning of scaffolds with multidimensional fiber architecture for hierarchical tissue engineering. ACS Appl Mater Inter. 2017;9(44):38187–200.

75. Ma PX, Zhang R. Synthetic nano-scale fibrous extracellular matrix. J Biomed Mater Res. 1999;46(1):60–72.

76. Hu J, Feng K, Liu X, Ma PX. Chondrogenic and osteogenic differentiations of human bone marrow-derived mesenchymal stem cells on a nanofibrous scaffold with designed pore network. Biomaterials. 2009;30(28):5061–7.

77. Gupte MJ, Swanson WB, Hu J, Jin X, Ma H, Zhang Z, Liu Z, Feng K, Feng G, Xiao G and others. Pore size directs bone marrow stromal cell fate and tissue regeneration in nanofibrous macroporous scaffolds by mediating vascularization. Acta Biomater. 2018;82:1–11.

78. He L, Liu B, Xipeng G, Xie G, Liao S, Quan D, Cai D, Lu J, Ramakrishna S. Microstructure and properties of nano-fibrous PCL-b-PLLA scaffolds for cartilage tissue engineering. Eur Cell Mater. 2009;18:63–74.

79. Prasopthum A, Shakesheff KM, Yang J. Direct three-dimensional printing of polymeric scaffolds with nanofibrous topography. Biofabrication. 2018;10(2):025002.

80. Mahapatra C, Kim JJ, Lee JH, Jin GZ, Knowles JC, Kim HW. Differential chondro- and osteo-stimulation in three-dimensional porous scaffolds with different topological surfaces provides a design strategy for biphasic osteochondral engineering. J Tissue Eng. 2019;10:2041731419826433.

81. Zhang S. Fabrication of novel biomaterials through molecular self-assembly. Nat Biotechnol. 2003;21(10):1171–8.

82. Lim EH, Sardinha JP, Myers S. Nanotechnology biomimetic cartilage regenerative scaffolds. Arch Plast Surg. 2014;41(3):231–40.
83. Shah RN, Shah NA, Del Rosario Lim MM, Hsieh C, Nuber G, Stupp SI. Supramolecular design of self-assembling nanofibers for cartilage regeneration. Proc Natl Acad Sci USA. 2010;107(8):3293–8.
84. Sun X, Yin H, Wang Y, Lu J, Shen X, Lu C, Tang H, Meng H, Yang S, Yu W and others. In situ articular cartilage regeneration through endogenous reparative cell homing using a functional bone marrow-specific scaffolding system. ACS Appl Mater Inter. 2018;10(45):38715–28.
85. Childs A, Hemraz UD, Castro NJ, Fenniri H, Zhang LG. Novel biologically-inspired rosette nanotube PLLA scaffolds for improving human mesenchymal stem cell chondrogenic differentiation. Biomed Mater. 2013;8(6):065003.
86. Park GE, Pattison MA, Park K, Webster TJ. Accelerated chondrocyte functions on NaOH-treated PLGA scaffolds. Biomaterials. 2005;26(16):3075–82.
87. Balasundaram G, Storey DM, Webster TJ. Novel nano-rough polymers for cartilage tissue engineering. Int J Nanomed. 2014;9:1845–53.
88. Wu YN, Law JB, He AY, Low HY, Hui JH, Lim CT, Yang Z, Lee EH. Substrate topography determines the fate of chondrogenesis from human mesenchymal stem cells resulting in specific cartilage phenotype formation. Nanomedicine. 2014;10(7):1507–16.
89. Cohen NP, Foster RJ, Mow VC. Composition and dynamics of articular cartilage: structure, function, and maintaining healthy state. J Orthop Sports Phys Ther. 1998;28(4):203–15.
90. Chahine NO, Collette NM, Thomas CB, Genetos DC, Loots GG. Nanocomposite scaffold for chondrocyte growth and cartilage tissue engineering: effects of carbon nanotube surface functionalization. Tissue Eng Part A. 2014;20(17–18):2305–15.
91. Boyer C, Figueiredo L, Pace R, Lesoeur J, Rouillon T, Visage CL, Tassin JF, Weiss P, Guicheux J, Rethore G. Laponite nanoparticle-associated silated hydroxypropylmethyl cellulose as an injectable reinforced interpenetrating network hydrogel for cartilage tissue engineering. Acta Biomater. 2018;65:112–22.
92. Chen J, An R, Han L, Wang X, Zhang Y, Shi L, Ran R. Tough hydrophobic association hydrogels with self-healing and reforming capabilities achieved by polymeric core-shell nanoparticles. Mater Sci Eng C Mater Biol Appl. 2019;99:460–7.
93. Naseri N, Deepa B, Mathew AP, Oksman K, Girandon L. Nanocellulose-based Interpenetrating Polymer Network (IPN) hydrogels for cartilage applications. Biomacromol. 2016;17(11):3714–23.
94. Wang C, Shen H, Tian Y, Xie Y, Li A, Ji L, Niu Z, Wu D, Qiu D. Bioactive nanoparticle-gelatin composite scaffold with mechanical performance comparable to cancellous bones. ACS Appl Mater Inter. 2014;6(15):13061–8.
95. Bellucci G, Seedhom BB. Mechanical behaviour of articular cartilage under tensile cyclic load. Rheumatology (Oxford). 2001;40(12):1337–45.
96. Racunica TL, Teichtahl AJ, Wang Y, Wluka AE, English DR, Giles GG, O'Sullivan R, Cicuttini FM. Effect of physical activity on articular knee joint structures in community-based adults. Arthritis Rheum. 2007;57(7):1261–8.
97. Andriacchi TP, Favre J, Erhart-Hledik JC, Chu CR. A systems view of risk factors for knee osteoarthritis reveals insights into the pathogenesis of the disease. Ann Biomed Eng. 2015;43(2):376–87.
98. Sanchez-Adams J, Leddy HA, McNulty AL, O'Conor CJ, Guilak F. The mechanobiology of articular cartilage: bearing the burden of osteoarthritis. Curr Rheumatol Rep. 2014;16(10):451.
99. Son B, Kim HD, Kim M, Kim JA, Lee J, Shin H, Hwang NS, Park TH. Physical stimuli-induced chondrogenic differentiation of mesenchymal stem cells using magnetic nanoparticles. Adv Healthc Mater. 2015;4(9):1339–47.
100. Zhang N, Lock J, Sallee A, Liu H. Magnetic nanocomposite hydrogel for potential cartilage tissue engineering: synthesis, characterization, and cytocompatibility with bone marrow derived mesenchymal stem cells. ACS Appl Mater Inter. 2015;7(37):20987–98.
101. Brady MA, Talvard L, Vella A, Ethier CR. Bio-inspired design of a magnetically active trilayered scaffold for cartilage tissue engineering. J Tissue Eng Regen Med. 2017;11(4):1298–302.

102. Rotherham M, El Haj AJ. Remote activation of the Wnt/beta-catenin signalling pathway using functionalised magnetic particles. PLoS ONE. 2015;10(3):e0121761.

103. Kanczler JM, Sura HS, Magnay J, Green D, Oreffo RO, Dobson JP, El Haj AJ. Controlled differentiation of human bone marrow stromal cells using magnetic nanoparticle technology. Tissue Eng Part A. 2010;16(10):3241–50.

104. Stolfa S, Skorvanek M, Stolfa P, Rosocha J, Vasko G, Sabo J. Effects of static magnetic field and pulsed electromagnetic field on viability of human chondrocytes in vitro. Physiol Res. 2007;56(Suppl 1):S45–9.

105. Amin HD, Brady MA, St-Pierre JP, Stevens MM, Overby DR, Ethier CR. Stimulation of chondrogenic differentiation of adult human bone marrow-derived stromal cells by a moderate-strength static magnetic field. Tissue Eng Part A. 2014;20(11–12):1612–20.

106. Go G, Han J, Zhen J, Zheng S, Yoo A, Jeon MJ, Park JO, Park S. A magnetically actuated microscaffold containing mesenchymal stem cells for articular cartilage repair. Adv Healthc Mater. 2017;6(13).

107. Urban JP, Hall AC, Gehl KA. Regulation of matrix synthesis rates by the ionic and osmotic environment of articular chondrocytes. J Cell Physiol. 1993;154(2):262–70.

108. Mobasheri A, Matta C, Uzieliene I, Budd E, Martin-Vasallo P, Bernotiene E. The chondrocyte channelome: a narrative review. Joint Bone Spine. 2019;86(1):29–35.

109. Xu J, Wang W, Clark CC, Brighton CT. Signal transduction in electrically stimulated articular chondrocytes involves translocation of extracellular calcium through voltage-gated channels. Osteoarthr Cartilage. 2009;17(3):397–405.

110. Khang D, Lu J, Yao C, Haberstroh KM, Webster TJ. The role of nanometer and submicron surface features on vascular and bone cell adhesion on titanium. Biomaterials. 2008;29(8):970–83.

111. Park JS, Yang HN, Woo DG, Chung HM, Park KH. In vitro and in vivo chondrogenesis of rabbit bone marrow-derived stromal cells in fibrin matrix mixed with growth factor loaded in nanoparticles. Tissue Eng Part A. 2009;15(8):2163–75.

112. Santo VE, Popa EG, Mano JF, Gomes ME, Reis RL. Natural assembly of platelet lysate-loaded nanocarriers into enriched 3D hydrogels for cartilage regeneration. Acta Biomater. 2015;19:56–65.

113. Lu H, Lv L, Dai Y, Wu G, Zhao H, Zhang F. Porous chitosan scaffolds with embedded hyaluronic acid/chitosan/plasmid-DNA nanoparticles encoding TGF-beta1 induce DNA controlled release, transfected chondrocytes, and promoted cell proliferation. PLoS ONE. 2013;8(7):e69950.

114. Mirza EH, Pan-Pan C, Wan Ibrahim WM, Djordjevic I, Pingguan-Murphy B. Chondroprotective effect of zinc oxide nanoparticles in conjunction with hypoxia on bovine cartilage-matrix synthesis. J Biomed Mater Res A. 2015;103(11):3554–63.

115. Albro MB, Nims RJ, Durney KM, Cigan AD, Shim JJ, Vunjak-Novakovic G, Hung CT, Ateshian GA. Heterogeneous engineered cartilage growth results from gradients of media-supplemented active TGF-beta and is ameliorated by the alternative supplementation of latent TGF-beta. Biomaterials. 2016;77:173–85.

116. Albro MB, Bergholt MS, St-Pierre JP, Vinals Guitart A, Zlotnick HM, Evita EG, Stevens MM. Raman spectroscopic imaging for quantification of depth-dependent and local heterogeneities in native and engineered cartilage. NPJ Regen Med. 2018;3:3.

117. Gibson M, Beachley V, Coburn J, Bandinelli PA, Mao HQ, Elisseeff J. Tissue extracellular matrix nanoparticle presentation in electrospun nanofibers. Biomed Res Int. 2014;2014:469120.

118. Li Y, Jiang T, Zheng L, Zhao J. Osteogenic differentiation of mesenchymal stem cells (MSCs) induced by three calcium phosphate ceramic (CaP) powders: a comparative study. Mater Sci Eng C Mater Biol Appl. 2017;80:296–300.

119. Kon E, Delcogliano M, Filardo G, Pressato D, Busacca M, Grigolo B, Desando G, Marcacci M. A novel nano-composite multi-layered biomaterial for treatment of osteochondral lesions: technique note and an early stability pilot clinical trial. Injury. 2010;41(7):693–701.

120. Amadori S, Torricelli P, Panzavolta S, Parrilli A, Fini M, Bigi A. Multi-layered scaffolds for osteochondral tissue engineering: in vitro response of co-cultured human mesenchymal stem cells. Macromol Biosci. 2015;15(11):1535–45.
121. Radhakrishnan J, Manigandan A, Chinnaswamy P, Subramanian A, Sethuraman S. Gradient nano-engineered in situ forming composite hydrogel for osteochondral regeneration. Biomaterials. 2018;162:82–98.
122. Li C, Armstrong JP, Pence IJ, Kit-Anan W, Puetzer JL, Correia Carreira S, Moore AC, Stevens MM. Glycosylated superparamagnetic nanoparticle gradients for osteochondral tissue engineering. Biomaterials. 2018;176:24–33.
123. Li C, Ouyang L, Pence IJ, Moore AC, Lin Y, Winter CW, Armstrong JPK, Stevens MM. Buoyancy-driven gradients for biomaterial fabrication and tissue engineering. Adv Mater. 2019:e1900291.
124. Rothenfluh DA, Bermudez H, O'Neil CP, Hubbell JA. Biofunctional polymer nanoparticles for intra-articular targeting and retention in cartilage. Nat Mater. 2008;7(3):248–54.
125. Bajpayee AG, Grodzinsky AJ. Cartilage-targeting drug delivery: can electrostatic interactions help? Nat Rev Rheumatol. 2017;13(3):183–93.

Chapter 6
Nanomaterials for Engineering the Treatment of Skin Wounds

Manuel Ahumada, Ying Wang and Walfre Franco

Abstract The skin is the largest organ in the human body; however, it is only a few millimeters thick. Among the main functions of the skin are to serve as a barrier for protection against physical and biological insults, as a thermal regulator to control internal temperatures, and as a sensor of physical stimulus that could lead to pleasant or harmful experiences. This highly-integrated sensory and regulatory armor is also capable of self-repair in response to injury, albeit the quality and extent of healing are determined by the skin condition and the type and size of the wound. In general, the wound healing process of skin comprehends four stages, which are hemostasis, inflammation, proliferation, and remodeling. These stages can be affected by internal physiological conditions and external environmental factors compromising the healing of the wound, for example, chronic wounds and bacterial infections. This chapter opens with an overview of the physiology of skin and skin wound healing. This overview is followed by a review of nanomaterial technologies and methods that have been investigated for the treatment of skin wounds. The chapter ends with an outlook of nanotechnology strategies for improving the treatment of skin wounds.

6.1 Introduction

Skin is the largest organ in the human body. The biology of the skin involves a myriad of cell types and structures that work together to enable key organ functions [1, 2]. We interact with our environment through the skin, which also protects us from the environment. Our skin plays an essential role in how we present ourselves to

M. Ahumada
Center for Applied Nanotechnology, Faculty of Sciences, Universidad Mayor, Huechuraba 8580745, RM, Chile

Y. Wang · W. Franco (✉)
Wellman Center for Photomedicine, Massachusetts General Hospital, Boston, MA 02114, USA
e-mail: wfranco@mgh.harvard.edu

Department of Dermatology, Harvard Medical School, Boston, MA 02114, USA

© Springer Nature Switzerland AG 2019
E. I. Alarcon and M. Ahumada (eds.), *Nanoengineering Materials for Biomedical Uses*,
https://doi.org/10.1007/978-3-030-31261-9_6

society—from tanning and tattoos to scars and wrinkles. When the skin is breached or wounded, the immediate organ response is to initiate a repair process to close the gap, and while many traditional and modern standard approaches to promote wound healing have proven effective in saving countless lives, novel approaches enabling healing of complex wounds and healing by remodeling—without scarring—are still missing.

6.2   The Basic Biology of Human Skin

The skin consists of two main layers: the top layer is called the epidermis, which constitutes an epithelium (tissue that covers an inner or outer surface), and the lower layer is called the dermis, which constitutes a connective tissue (Fig. 6.1).

The epidermis is comprised mainly by a multilayered epithelium, the interfollicular epidermis, and adnexal structures. The adnexal structures, like hair follicles, sebaceous glands, and sweat glands, provide function to the skin [1]. The thickness of the epidermis and the distribution of adnexal structures vary across body sites. The main cell type in the epidermis is the keratinocyte. Among the key functions of the

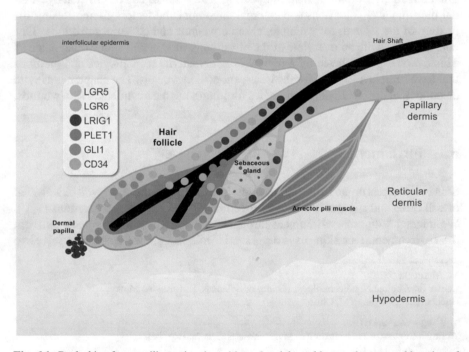

Fig. 6.1 Back skin of mouse illustrating the epidermal and dermal layers; the type and location of different stem cell populations (LGR6 and LRIG1 are expressed in the follicle isthmus and LGR5 and CD34 in the bulge); and, the specialized mesenchymal cells of the dermal papilla and arrector pili muscle. Adapted from Watt [1]

epidermis are the formation of a shielding interface with the outside environment, corporeal thermal regulation by hairs and sweat, sensory perception of heat, pain, pressure by nerves, and lubrication of the skin with lipids. Most of these functions stem from nondividing and terminally differentiated keratinocyte cells, which are in constant turn over. Keratinocyte cells are replenished through a variety of stem cell populations that are located in different skin sites, such as hair follicles (Fig. 6.1) [3]. Under normal physiological conditions, each stem cell cluster produces a subset of differentiated epidermal cells; however, most stem cells are able to contribute to the entire variety of differentiated epidermal lineages when the cells are relocated, or the skin is wounded.

The dermis consists of three layers: the upper layer in contact with the epidermis is the papillary dermis, the middle layer is the reticular dermis, and the deepest layer is termed the hypodermis. A basement membrane separates the dermis from the epidermis. This membrane is extracellular matrix characterized by an abundance of type IV collagen and laminin. The main cell type in the dermis is the fibroblast, and its density is higher in the papillary dermis. The reticular dermis is rich in fibrillar collagen, and the hypodermis is characterized by the presence of white adipocytes. Two mesenchymal structures important for function in the dermis are the dermal papilla—a population of cells that regulates the hair cycle and is located at the base of the follicle—and the arrector pili muscle, which a smooth muscle that upon contraction straights the hair follicles [1].

In addition to keratinocytes and fibroblasts, other important cell types, which either reside in the skin or circulate through the skin, are melanocytes, innate and adaptive immune system cells, peripheral nervous system cells, and cells of blood vessels [4–6].

6.3 When the Skin Is Wounded

The capacity of the skin to heal after an injury is vital for our survival, and this capacity is often limited, compromised, and disrupted in a spectrum of conditions and disorders. The process of skin wound healing is complex, it requires a highly coordinated response by a multitude of cells, like, immune cells, hematopoietic cells (immature cells that can differentiate into all types of blood cells), and resident cells of the skin (keratinocytes, fibroblasts). Skin wound healing is typically divided into four stages that overlap: hemostasis, inflammation, proliferation, and remodeling. In each stage, multiple key molecular, cellular, and physiologic events take place, and these events are coordinated primarily by signaling between immunologic, hematopoietic and resident skin cells. These stages are schematically represented in Fig. 6.2 and have been reviewed in detail in [7].

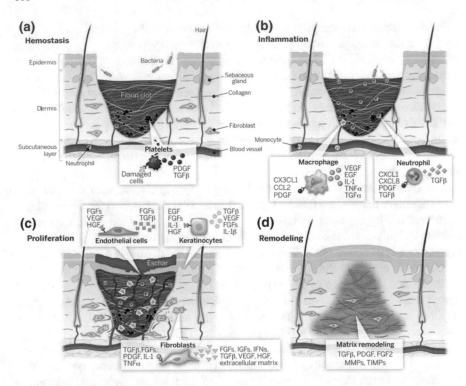

Fig. 6.2 Overlapping stages of skin wound healing: **a** hemostasis, **b** inflammation, **c** proliferation, **d** remodeling. A multitude of fundamental molecular and cellular processes coordinated by a myriad of secreted factors occur at each stage. The schematic illustrates representative factors from each stage. Reproduced with permission from Sun et al. [2]

6.3.1 Skin Wound Healing

In humans, immediately after the skin is wounded, hemostasis begins. Multiple responses are activated to stop blood loss: local vascular smooth muscle cells constrict vessels, reducing the blood flow; platelets and coagulation cascade factors produce fibrin, forming a hemostatic clot and a scaffold for the infiltration of leukocytes, keratinocytes, fibroblasts, and other cells into the wound [8]. The inflammatory stage begins within hours from the time of injury. Inflammation is driven by mediators derived from platelets, by-products from bacteria, and chemoattractants secreted in the wound. The first cell type that infiltrates the wound site is the neutrophil in order to kill bacteria and degrade matrix proteins that are damaged [9]. Within 24 h, monocytes cells infiltrate the wound and differentiate into macrophages, which kill microbes, remove tissue debris, eliminate neutrophils, and enable angiogenesis and tissue granulation [10].

During the proliferation stage, new cells populate the wound and a multitude of growth factors and chemokines that induce cell migration and proliferation, and matrix formation are released: platelet-derived growth factor (PDGF), fibroblast growth factors (FGFs), vascular endothelial growth factor (VEGF), and transforming growth factor -α and -β (TGF-α and TGF-β), among others. In addition, keratinocyte cells are released from the wound edges and stem cell reservoirs located in the bulge and isthmus of hair follicles and the interfollicular epidermis. Keratinocytes proliferate and migrate to close the wound by coverage. Upon contact inhibition, keratinocytes undergo vertical stratification and differentiation to reinstate all the layers of the epidermal barrier [11, 12]. Angiogenesis, in synchrony with epidermal repair, is activated by multiple growth factors, for example, VEGF and FGF. The combination of new blood vessels, fibroblast and macrophage cells, and matrix proteins forms granulation tissue, which is the soft and pink tissue that forms at the bottom of a healing skin wound. At the end of the proliferative stage, fibroblast cells differentiate into contractile myofibroblast cells that pull together the wound edges [13].

During the remodeling stage, cells from previous stages are removed and collagen deposition in the dermis transitions from type III to type I collagen. Collagen remodeling involves matrix metalloproteinases, and its synthesis increases the tensile strength of the wounded skin, which recovers approximately 40% of its original strength in 4 weeks and about 70% in 1 year in normal healing conditions [14]. Altered collagen synthesis during this stage leads to scar formation [15]. Furthermore, poor or pathologic wound healing originates from failure to initiate, terminate, or regulate any particular wound healing stage. Examples of poor and pathologic wound healing outcomes are pyogenic granulomas (excessive formation of granulation tissue), hypertrophic scars and keloids (accumulation of excess fibrotic components), and chronic ulcers (prolonged inflammation, poor vascularization, and inability to form a new epithelium).

6.4 Nanomaterials for Engineering Skin Wound Healing

The skin endures and heals injuries throughout our lives. And healing is influenced by a wide variety of factors that affect skin wounding and the speed and quality of healing. These factors include a number of common conditions, diseases, and treatments, underscoring the far-reaching relevance of skin wound healing to medicine, public health, and the global burden of disease. Surgical incisions, thermal burns, and chronic ulcers are among the conditions that significantly burden (or delay) wound healing [2]. Current clinical approaches to promote wound healing have proven effective in saving lives; however, novel approaches enabling healing of complex wounds and healing by remodeling—without scarring—are still missing. Nanomedicine has demonstrated its potential to advance healthcare by developing nanoscale methods and technology to address a broad range of clinical limitations in drug delivery, biosensing, imaging, tissue regeneration, diagnostics, cancer, and other diseases treatments [16–18]. Nanomaterials have been extensively investigated for improved skin wound-healing and have demonstrated, mostly in animal models, their ability in regulating skin wound healing at different stages.

Next, we present an illustrative summary of the type of nanomaterials that have been tested in animal models of skin wounds, leaving out cell studies for succinctness and including only a few human studies because there are not many. A variety of nanomaterial frameworks—polymer nanoparticles, dendrimers, liposomes, metals, ceramics, fullerenes, nanotubes, nanoemulsions, nanopores, quantum dots—have been utilized to address specific healing limitations [19]. The unique properties of nanomaterials and their potential applications for the treatment of tissues other than skin—bone, bladder, cartilage, neural, vascular, etc.—are reviewed in [16, 20–22]. Excellent reviews about nanomaterials in skin wound healing can be found in [23–25].

6.4.1 Carbon

Carbon nanomaterials have been used for imaging, drug delivery, and gene delivery, among other applications [26, 27]. This type of nanomaterials includes fullerenes, carbon nanohorns, carbon nanotubes, and graphene. Fullerene is a powerful antioxidant capable of scavenging ROS and reactive nitrogen species and, consequently, of modulating inflammatory and proliferative processes [28]. In combination with light, fullerene has also been used to rescue mice with wounds infected with pathogenic gram-negative bacteria [28]. The biocompatibility of carbon nanomaterials remains controversial [26], and most studies in the literature are limited to in vitro cell studies [29–31].

6.4.2 Ceramics

Ceramic nanomaterials based on silica, calcium salts and hydroxyapatite have been investigated as nanoparticles for applications in wound healing. Intravenously administered calcium-based nanoparticles, synthesized using $CaCl_2$ with β-glycerolphosphate, were used to modulate local calcium levels and calcium homeostasis in open wounds in mice. These nanoparticles decreased the size of the wound by contraction, which was attributed to the release of ionized calcium into the wound bed as the pH-sensitive nanoparticles dissolved in the acidic wound microenvironment [32]. Variations in local pH levels in the wound bed are known to promote or inhibit bacteria growth, regulate enzyme prevalence, and change oxygen supply [33]. A similar approach utilized nitric NO-releasing nanoparticles, which were synthesized by means of combining chitosan, glucose, tetramethyl orthosilicate, sodium nitrite, and polyethylene glycol. In clean and infected wounds in mice, the wounds treated with NO-NPs presented reduced inflammation, increased collagen deposition, and increased blood vessel formation. NO stimulated migration and proliferation of fibroblasts and synthesis of collagen type III [34, 35].

6.4.3 Lipids

Nanoparticle treatment approaches for accelerating wound closure have been developed using liposomes [36, 37], and tested in porcine models. Lipids obtained from the cell membranes of rabbit red blood cells were combined with α-gal epitopes to make submicroscopic liposomes (Gll-Phl-Chol-α-gal). Anti-Gal is an antibody that interacts specifically with a carbohydrate antigen, the α-gal epitope (Galα1-3Galβ1-4GlcNAc-R), on glycolipids and glycoproteins in a wound and activates the complement system. Among the products generated from this complement activation, there are complement cleavage chemotactic factors, such as C5a and C3a. The increment in local concentrations of chemotactic factors within the wound stimulated rapid recruitment and migration of macrophages to the wound site [36, 37]. α-Gal nanoparticles binding to the membrane receptors of macrophages were able to activate the production of cytokines and accelerate healing and wound closure, which was demonstrated in pig wounds treated topically with nanoparticles deposited as a thin film [36].

A different approach incorporated plant derived compounds quercetin and curcumin in liposomes. The drug and carrier efficacy were evaluated by in vitro skin distribution and in vivo ability to reduce oxidative inflammation and neutrophil infiltration in mice subjected to 12-O-tetradecanoylphorbol-13-acetate (TPA, a potent tumor promoter). These plant-based drugs present antioxidant and anti-inflammatory properties and were able to inhibit the onset of skin wounds during the application of TP [38]. A different study utilized phospholipid bilayer vesicles loaded with hemoglobin and coated with polyethylene glycol (HbVs) to improve the oxygenation of ischemic skin wounds in mice models [39]. Relative to controls, tissue survival improved by 24% and the wound-healing rate increased twofold. Immunohistochemical analysis showed a higher density of capillaries and a higher expression of endothelial NO synthase 3 in the wounds treated with HbVs. A different study investigated liposome-encapsulated hemoglobin with high O_2 affinity for the treatment of full-thickness dorsal wounds in mice [40]. The treatment significantly accelerated granulation, increased epithelial thickness, suppressed early granulocyte infiltration, and increased Ki67 expression.

Another approach synthesized adenosine triphosphate (ATP)-vesicles and tested them as a topical nonionic cream for improved wound healing. ATP-vesicles constructed using phospholipids, trehalose, and a liposomal transfection reagent, were combined with a nonionic commercial cream moisturizer. For both non-ischemic and ischemic wounds, the ATP-vesicle provided a source of energy for survival of cells and improved granulation and re-epithelialization in diabetic wounds in rabbits [41].

Solid lipid nanoparticles (SLNs) and nanostructured lipid carriers (NLCs) are two major types of lipid-based nanoparticles. SLNs advantages are good release profiles and targeted drug delivery with excellent physical stability and, consequently, have overcome the limitations of other colloidal carriers, like emulsions, liposomes, and polymeric nanoparticles. Topical administration of SLNPs and NLCs loaded with rhEGF improved re-epithelialization and restoration of the inflammatory process in diabetic wounds in mice [42]. A different approach developed silica (SiO_2) nanoparticles as an alternative to conventional wound closure methods, such as sutures and adhesives like Dermabond (2-octyl cyano-acrylate). SiO_2 nanoparticles synthesized by the Stöber method were applied to full-thickness dorsal skin wounds. The wound edges were maintained in contact manually for less than one minute, which is the time that took to close the wound. Macroscopic analysis shows the absence of pathological inflammation or necrosis [43]. A class of biomaterials that have been successfully implanted in millions of patients worldwide to repair bone and dental defects are bioactive glasses, primarily 45S5 glass compositions [44]. Many other bioactive glass compositions have been proposed for applications like soft tissue repair and drug delivery. A bioactive glass-based nanoformulation was applied to full-thickness wounds in a diabetic rat model. This nanoformulation promoted the proliferation of fibroblasts and deposition of granulation tissue while stimulating the production of growth factors such as VEGF and FGF2 [45].

6.4.4 Metal

A variety of metal nanoparticles has been investigated and developed for wound healing applications. Colloidal solutions of silver (Ag) nanoparticles possess a remarkably wide spectrum of antimicrobial properties, effectively reducing or preventing wound infections caused by a broad range of microbes [46, 47]. AgNPs are also effective against fungi, yeast, and viruses [48]. The origin and antimicrobial activity of these particles, including the actual contribution of ionic silver to the antimicrobial activity, is an ongoing discussion [49, 50]. However, it is well established that adequate surfaces are required for producing biologically active nanoparticles [51–58]. As a function of size and concentration, AgNPs have shown anti-inflammatory properties and the ability to minimize ROS production and improve the tensile strength of skin by modulating collagen alignment [59–61]. The production of stable and non-toxic nanosilver under physiological conditions constitutes a development approach for the translation of silver nanocomposites to the clinic. Cumulative data to date highlights the fundamental role of bio-inspired protecting agents to produce stable nanosilver structures that do not have toxic side effects on primary human cells and mice, i.e., protective agents like collagen and the thiol-modified LL37 antimicrobial peptide [62]. In situ preparation of fibers with AgNPs utilizing electrospinning was developed for use as a wound dressing. Nanofibrous membranes enabled the continuous release of Ag ions, which resulted in broad-spectrum antimicrobial activity against *Staphylococcus aureus* and *Escherichia coli*. These antibacterial nanofibrous membranes were able to reduce the inflammatory response and accelerate wound healing in Wistar rats [63].

In addition to silver, iron oxide, copper, and gold have been investigated for wound healing applications. Thrombin-conjugated γ-Fe_2O_3 nanoparticles accelerated the closure of incisional wounds in rats, improving skin tensile strength and reducing stitch-induced scarring [64]. Copper nanoparticles within a methylcellulose-based ointment were shown to induce pro-inflammatory mediators that increased blood vessel formation in full-thickness wounds in mice skin [65]. Small-sized gold nano-materials are stable and nontoxic, which makes them an attractive platform for development of biofunctional nanomaterials. Photoluminescent gold nanodots were constructed using etching and codeposition of hybridized ligands, an antimicrobial peptide (surfactant), and 1-dodecanethiol, on gold nanoparticles. In rats, wounds infected with Methicillin-Resistant S. aureus showed faster healing, improved epithelialization, and higher collagen deposition when photoluminescent gold nanodots are used as a dressing material [66].

Metals of the lanthanide group have also been investigated in wound healing [67]. Nanoceria (cerium oxide nanoparticles) is an ROS scavenger material that stabilizes HIF-1α expression stimulating angiogenesis via upregulation of VEGF and modulating oxygen levels in the wound [24, 68]. Nanoceria/PU/cellulose acetate fibers applied to wounds exhibited anti-bacterial properties due to the release of free cerium ions [69].

6.4.5   Polymers

Polymeric nanomaterials have been used for drug delivery, imaging and sensing applications [70]. In wound healing, polymeric nanomaterials are often combined with wound dressings [71, 72]. Poly (lactide-co-glycolide) (PLGA), polycaprolactone (PCL) and polyethylene glycol (PEG) have been used to develop nanomaterials for improving wound closure outcomes in normal and infected wounds. PLGA nanoparticles loaded with curcumin reduced the inflammatory response, accelerated re-epithelialization and improved the formation of granulation tissue in mice wounds [73]. Curcumin, an organic molecule found in the spice turmeric, exhibits anti-inflammatory, antioxidant and bactericidal properties [74, 75]. PLGA nanoparticles loaded with recombinant human epidermal growth factor (rhEGF) increased the healing rate of full-thickness wounds in diabetic rats [76]. The release of rhEGH was sustained for 24 h enhancing fibroblast proliferation. The application of PCL-PEG nanoparticles loaded with hypericin (Hy) to wounds infected with methicillin-resistant S. aureus resulted in reduced expression of TNF—tumor necrosis factor, a cytokine involved in systemic inflammation—improved epithelialization and collagen synthesis in a rat model [77].

Biodegradable poly(b-amino esters) (PBAEs) and copolymers of maleic acid have been exploited to develop gene delivery and drug release systems for wound-healing applications [78, 79]. PBAEs nanoparticles carrying the sonic hedgehog gene promoted angiogenesis and tissue regeneration by activating angiogenic signaling pathways in mice wound model [80]. Chitosan, an organic polysaccharide exhibiting biocompatibility, biodegradability, mucoadhesivity and anti-infection activity, has been extensively investigated for biomedical applications [81]. Chitosan nanoparticles have shown significant bactericidal effects on different types of bacteria without cell toxicity on mouse fibroblast cells [82]. The N-acetyl glucosamine in chitosan is also present in the dermal connective tissue in elastin, which is a highly-elastic structural and cell-signaling protein [83]. Synthesized chimeric nanoparticles, formed via spontaneous self-assembling of elastin-like peptides (ELP) and loaded with keratinocytes growth factors, improved re-epithelialization and granulation in diabetic mice wounds [84].

Dendrimers, like polyamidoamine (PAMAM), have been used for cellular delivery of plasmid DNA by forming a stable complex with limited degradation. An arginine-grafted cationic dendrimer (PAM) was used to deliver minicircle plasmid DNA encoding VEGF into diabetic wounds in mice. This polycomplex PAM-RG4) consists of a highly ordered PAMAM dendrimer backbone. Relative to control wounds, PAM-RG4 improved the proliferation of basal cells and the deposition of collagen, and it also reduced the formation of immature blood vessels. Diabetic skin wounds healed within 6 days displaying a well-ordered dermal structure [85]. The structure of dendrimers exhibits internal cavities and surface channels, making them suitable to accommodate small molecules and carry high drug loads that can be used for the treatment of complex wounds. Dendrimers have also been used in the development of nanotechnology for delivering short RNA molecules to wounds. RNA interference (RNAi) is a biological process in which small interfering RNAs (siRNAs) silence gene expression by degrading targeted mRNA molecules. Sirnaomics Inc., is developing nanoparticle formulations containing siRNA for reducing the expression of TGFβ1 and Cox-2 (cyclooxygenase-2), which are implicated in many tissue inflammation and fibrosis processes. Among other applications, Sirnaomics Inc., is testing formulations for the treatment of hypertrophic scars in phase II clinical trials.

6.5   Emerging Technologies

As illustrated in the previous section, the vast majority of the nanomaterial applications investigated for skin wound healing consist in utilizing them either to modulate different wound healing processes by direct interaction with the wound microenvironment or to deliver cargo that modulates the wound healing process. Alternative emerging technologies utilize nanomaterials in synergy with novel technologies in other fields [86].

Photosensitized crosslinking of proteins is being developed for medical treatments and many potential benefits have already been demonstrated in preclinical and clinical studies. Applications include sealing wounds, reattaching severed tissues, stiffening and strengthening tissues, decreasing inflammatory responses, and bioengineering tissues [87–90]. These treatments rely on light-initiated formation of covalent crosslinks between proteins on the surfaces of two tissues or between proteins within a tissue. Two dyes, rose Bengal (RB^{2-}) and riboflavin-5-phosphate ($R5P^{2-}$), has been used almost exclusively for medical applications of photosensitized protein crosslinking. The treatment procedure for photo-crosslinking tissue proteins is simple. In fact, the simplicity of this technique is one of its clinical advantages. An aqueous solution of the dye typically is applied to the tissue, which is then exposed for a few minutes to light wavelengths absorbed by the dye. Crosslinking occurs during the irradiation and is followed by healing processes. The light-induced effects appear to result largely from crosslinks in collagen, the major connective tissue protein, with possible crosslinking to the proteoglycans that surround collagen fibrils and to other proteins. This technique has been regarded as an effective alternative to stapling or traditional suturing. Absorption and scattering of light in tissue results from fundamental light–matter interactions and have enabled a variety of powerful optical techniques for therapy and imaging [91]. However, these interactions are also problematic as they limit the penetration of light in tissues. A poly(allylamine) (PAAm)-modified upconversion nanoparticle/hyaluronate–rose bengal (UCNP/PAAm/HA-RB) conjugate complex was developed for photochemical bonding of deep tissue with near-infrared (NIR) light illumination [92]. In a mice study, the UCNP/PAAm/HA-RB conjugate complex was efficiently delivered into deep tissue and accelerated tissue bonding upon NIR light illumination. HA in the outer layer of the complex facilitated the penetration of RB into the collagen layer of the dermis.

Skin grafting requires removing a layer of skin tissue from a donor site to transfer it to the wounded site. Skin grafts can be classified as split-thickness skin grafts (STSGs), consisting of the epidermis and the upper part of the dermis, or full-thickness skin grafts (FTSGs), consisting of the epidermis and the full-thickness dermis. A major limitation of this procedure is scarring at the donor site and, in STSGs, the absence of the deep dermal adnexal structures that give function to skin, such as hair follicles, sweat glands, and sebaceous glands. FTSGs require to create a full-thickness wound at the donor site and a good vascular bed for survival at the wounded site. Therefore, standard clinical outcomes are scarring at the donor and wounded site, and lack of skin function at the wounded site. Innovations in surgical grafting techniques have advanced skin grafting, but grafting methods are still limited [2, 93]. In mice and porcine models, a new approach that grafts thousands of full-thickness skin micrografts have been shown to eliminate scarring at the donor site while transferring the deep adnexal structure to the wound site [94, 95]. Again, the simplicity of this technique is one of its clinical advantages [96, 97]. This fractional grafting approach places the micrografts randomly in the wounded site, that is, micrografts are not oriented. Current development efforts are geared toward developing nanomaterials that could serve as functional scaffolds to "copy skin" by enabling controlling the spacing and orientation of micrografts, which presumably would result in faster healing and functional skin resembling the original skin, and manipulation of this hybrid construct (autologous tissue in a synthetic biomaterial) by clinicians. The concept is illustrated in Fig. 6.3. The flexibility and broad reach

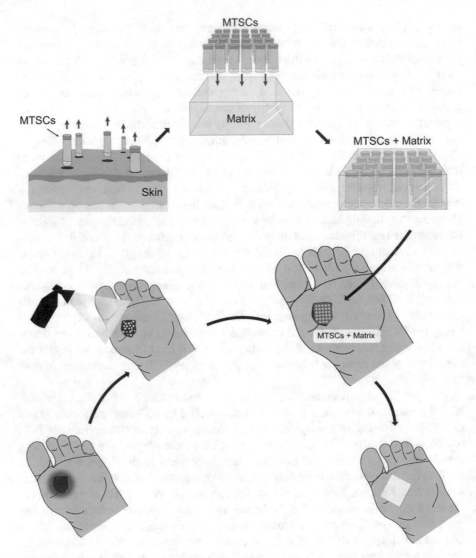

Fig. 6.3 Full-thickness microscopic skin tissue columns (MSTCs) are harvested from the healthy donor site and placed into a scaffold matrix to assemble a hybrid skin construct for wound repair. Nanomaterials enable functionalization of the matrix for different wound environments and modifying the wound to establish favorable conditions, for example, the combination with a silver nanoparticles spray for infected wounds

of nanomaterials in wound healing could enable a fractional skin grafting platform with the capability of customizing scaffolds for different wound environments; for example, in infected wounds by functionalizing the scaffold to kill bacteria while preserving the micrografts.

6.6 Summary and Conclusions

Wound healing interventions aim to repair or promote the repair of the structure and function of organs, tissues, and cells. When the skin is wounded, repair implies restoration of all functional components, including hair follicles, sweat glands, and nerves in clean and impaired healing conditions. Although there is no perfect repair method currently available for complex wounds, rapid developments in understanding skin development and wound repair, together with advances in related fields like stem cell, tissue bioengineering, and nanomedicine, provide hope that such methods represent a tractable goal in the future. Nanomaterials have a high surface area to mass ratio that enables favorable interactions with wound environment and constituents. Cells naturally interact with the extracellular matrix which can be better modulated by nanoscale materials. Nanomaterials will have a critical role to play in the engineering of novel clinical strategies that enable healing of large and complex wounds by remodeling without scarring.

Acknowledgements We would like to thank Dr. Rox R. Anderson for his support. All authors have read and approved this final version. Dr. Ahumada acknowledges the support of CONICYT-FONDECYT Iniciación (grant #11180616). Illustrating support was provided by Yanjie Jack Guo.

Disclosure All authors have read and approved this final version.

References

1. Watt FM. Mammalian skin cell biology: at the interface between laboratory and clinic. Science. 2014;346(6212):937–40.
2. Sun BK, Siprashvili Z, Khavari PA. Advances in skin grafting and treatment of cutaneous wounds. Science. 2014;346(6212):941–5.
3. Schepeler T, Page ME, Jensen KB. Heterogeneity and plasticity of epidermal stem cells. Development. 2014;141(13):2559–67.
4. Belkaid Y, Segre JA. Dialogue between skin microbiota and immunity. Science. 2014;346(6212):954–9.
5. Lo JA, Fisher DE. The melanoma revolution: from UV carcinogenesis to a new era in therapeutics. Science. 2014;346(6212):945–9.
6. Zimmerman A, Bai L, Ginty DD. The gentle touch receptors of mammalian skin. Science. 2014;346(6212):950–4.
7. Gurtner GC, Werner S, Barrandon Y, Longaker MT. Wound repair and regeneration. Nature. 2008;453(7193):314–21.

8. Clark RAF. Fibrin and wound healing. In: Nieuwenhuizen W, Mosesson MW, DeMaat MPM, editors. Fibrinogen, vol. 936. Annals of the New York Academy of Sciences; 2001. p. 355–67.
9. Ross R, Odland G. Human wound repair: II. Inflammatory cells epithelial-mesenchymal interrelations and fibrogenesis. J Cell Biol. 1968;39(1):152–68.
10. Koh TJ, DiPietro LA. Inflammation and wound healing: the role of the macrophage. Expert Rev Mol Med. 2011;13:e23.
11. Blanpain C, Fuchs E. Stem cell plasticity. Plasticity of epithelial stem cells in tissue regeneration. Science. 2014;344(6189):1243.
12. Ito M, Liu YP, Yang ZX, Nguyen J, Liang F, Morris RJ, Cotsarelis G. Stem cells in the hair follicle bulge contribute to wound repair but not to homeostasis of the epidermis. Nat Med. 2005;11(12):1351–4.
13. Tomasek JJ, Gabbiani G, Hinz B, Chaponnier C, Brown RA. Myofibroblasts and mechano-regulation of connective tissue remodelling. Nat Rev Mol Cell Biol. 2002;3(5):349–63.
14. Levenson SM, Geever EF, Crowley LV, Oates JF, Berard CW, Rosen H. Healing of rat skin wounds. Ann Surg. 1965;161(2):293.
15. Gill SE, Parks WC. Metalloproteinases and their inhibitors: regulators of wound healing. Int J Biochem Cell Biol. 2008;40(6–7):1334–47.
16. Tocco I, Zavan B, Bassetto F, Vindigni V. Nanotechnology-based therapies for skin wound regeneration. J Nanomater. 2012;11.
17. Etheridge ML, Campbell SA, Erdman AG, Haynes CL, Wolf SM, McCullough J. The big picture on nanomedicine: the state of investigational and approved nanomedicine products. Nanomedicine. 2013;9(1):1–14.
18. Ryan SM, Brayden DJ. Progress in the delivery of nanoparticle constructs: towards clinical translation. Curr Opin Pharmacol. 2014;18:120–8.
19. Athar M, Das AJ. Therapeutic nanoparticles: State-of-the-art of nanomedicine. Adv Mater Rev. 2014;1(1):25–37.
20. Cortivo R, Vindigni V, Iacobellis L, Abatangelo G, Pinton P, Zavan B. Nanoscale particle therapies for wounds and ulcers. Nanomedicine. 2010;5(4):641–56.
21. Obregon R, Ramon-Azcon J, Ahadian S, Shiku H, Bae H, Ramalingam M, Matsue T. The use of microtechnology and nanotechnology in fabricating vascularized tissues. J Nanosci Nanotechnol. 2014;14(1):487–500.
22. Zhang LJ, Webster TJ. Nanotechnology and nanomaterials: promises for improved tissue regeneration. Nano Today. 2009;4(1):66–80.
23. Chakrabarti S, Chattopadhyay P, Islam J, Ray S, Raju PS, Mazumder B. Aspects of nanomaterials in wound healing. Curr Drug Deliv. 2019;16(1):26–41.
24. Kalashnikova I, Das S, Seal S. Nanomaterials for wound healing: scope and advancement. Nanomedicine. 2015;10(16):2593–612.
25. Mordorski B, Prow T. Nanomaterials for wound healing. Curr Dermatol Rep. 2016;5(4):278–86.
26. Zhang YB, Petibone D, Xu Y, Mahmood M, Karmakar A, Casciano D, Ali S, Biris AS. Toxicity and efficacy of carbon nanotubes and graphene: the utility of carbon-based nanoparticles in nanomedicine. Drug Metab Rev. 2014;46(2):232–46.
27. Zhou Z. Liposome formulation of fullerene-based molecular diagnostic and therapeutic agents. Pharmaceutics. 2013;5(4):525–41.
28. Lu Z, Dai T, Huang L, Kurup DB, Tegos GP, Jahnke A, Wharton T, Hamblin MR. Photodynamic therapy with a cationic functionalized fullerene rescues mice from fatal wound infections. Nanomedicine. 2010;5(10):1525–33.
29. Gao J, Wang HL, Iyer R. Suppression of proinflammatory cytokines in functionalized fullerene-exposed dermal keratinocytes. J Nanomater. 2010.
30. Ryoo SR, Kim YK, Kim MH, Min DH. Behaviors of NIH-3T3 fibroblasts on graphene/carbon nanotubes: proliferation, focal adhesion, and gene transfection studies. ACS Nano. 2010;4(11):6587–98.
31. Zhang YY, Wang B, Meng XA, Sun GQ, Gao CY. Influences of acid-treated multiwalled carbon nanotubes on fibroblasts: proliferation, adhesion, migration, and wound healing. Ann Biomed Eng. 2011;39(1):414–26.

32. Kawai K, Larson BJ, Ishise H, Carre AL, Nishimoto S, Longaker M, Lorenz HP. Calcium-based nanoparticles accelerate skin wound healing. PLOS One. 2011;6(11).
33. Schneider LA, Korber A, Grabbe S, Dissemond J. Influence of pH on wound-healing: a new perspective for wound-therapy? Arch Dermatol Res. 2007;298(9):413–20.
34. Blecher K, Martinez LR, Tuckman-Vernon C, Nacharaju P, Schairer D, Chouake J, Friedman JM, Alfieri A, Guha C, Nosanchuk JD and others. Nitric oxide-releasing nanoparticles accelerate wound healing in NOD-SCID mice. Nanomedicine. 2012;8(8):1364–71.
35. Han G, Nguyen LN, Macherla C, Chi YL, Friedman JM, Nosanchuk JD, Martinez LR. Nitric oxide-releasing nanoparticles accelerate wound healing by promoting fibroblast migration and collagen deposition. Am J Pathol. 2012;180(4):1465–73.
36. Hurwitz ZM, Ignotz R, Lalikos JF, Galili U. Accelerated porcine wound healing after treatment with alpha-gal nanoparticles. Plast Reconstr Surg. 2012;129(2):242E–51E.
37. Wigglesworth KM, Racki WJ, Mishra R, Szomolanyi-Tsuda E, Greiner DL, Galili U. Rapid recruitment and activation of macrophages by anti-gal/alpha-gal liposome interaction accelerates wound healing. J Immunol. 2011;186(7):4422–32.
38. Castangia I, Nacher A, Caddeo C, Valenti D, Fadda AM, Diez-Sales O, Ruiz-Sauri A, Manconi M. Fabrication of quercetin and curcumin bionanovesicles for the prevention and rapid regeneration of full-thickness skin defects on mice. Acta Biomater. 2014;10(3):1292–300.
39. Plock JA, Rafatmehr N, Sinovcic D, Schnider J, Sakai H, Tsuchida E, Banic A, Erni D. Hemoglobin vesicles improve wound healing and tissue survival in critically ischemic skin in mice. Am J Physiol Heart Circ Physiol. 2009;297(3):H905–10.
40. Fukui T, Kawaguchi AT, Takekoshi S, Miyasaka M, Tanaka R. Liposome-encapsulated hemoglobin accelerates skin wound healing in mice. Artif Organs. 2012;36(2):161–9.
41. Wang JP, Wan R, Mo YQ, Li M, Zhang QW, Chien SF. Intracellular delivery of adenosine triphosphate enhanced healing process in full-thickness skin wounds in diabetic rabbits. Am J Surg. 2010;199(6):823–32.
42. Gainza G, Pastor M, Aguirre JJ, Villullas S, Pedraz JL, Hernandez RM, Igartua M. A novel strategy for the treatment of chronic wounds based on the topical administration of rhEGF-loaded lipid nanoparticles: In vitro bioactivity and in vivo effectiveness in healing-impaired db/db mice. J Control Release. 2014;185:51–61.
43. Meddahi-Pelle A, Legrand A, Marcellan A, Louedec L, Letourneur D, Leibler L. Organ repair, hemostasis, and in vivo bonding of medical devices by aqueous solutions of nanoparticles. Angew Chem. 2014;53(25):6369–73.
44. Baino F, Hamzehlou S, Kargozar S. Bioactive glasses: where are we and where are we going? J Funct Biomat. 2018;9(1).
45. Lin C, Mao C, Zhang JJ, Li YL, Chen XF. Healing effect of bioactive glass ointment on full-thickness skin wounds. Biomed Mat. 2012;7(4).
46. Eckhardt S, Brunetto PS, Gagnon J, Priebe M, Giese B, Fromm KM. Nanobio silver: its interactions with peptides and bacteria, and its uses in medicine. Chem Rev. 2013;113(7):4708–54.
47. Alarcon EI, Griffith M, Udekwu KI. Silver nanoparticle applications. New York: Springer; 2015.
48. Lara HH, Garza-Trevino EN, Ixtepan-Turrent L, Singh DK. Silver nanoparticles are broad-spectrum bactericidal and virucidal compounds. J Nanobiotechnol. 2011;9.
49. Xiu ZM, Ma J, Alvarez PJJ. Differential effect of common ligands and molecular oxygen on antimicrobial activity of silver nanoparticles versus silver ions. Environ Sci Technol. 2011;45(20):9003–8.
50. Xiu ZM, Zhang QB, Puppala HL, Colvin VL, Alvarez PJJ. Negligible particle-specific antibacterial activity of silver nanoparticles. Nano Lett. 2012;12(8):4271–5.
51. Ahumada M, McLaughlin S, Pacioni NL, Alarcon EI. Spherical silver nanoparticles in the detection of thermally denatured collagens. Anal Bioanal Chem. 2016;408(8):1993–6.
52. Alarcon EI, Udekwu K, Skog M, Pacioni NL, Stamplecoskie KG, Gonzalez-Bejar M, Polisetti N, Wickham A, Richter-Dahlfors A, Griffith M and others. The biocompatibility and antibacterial properties of collagen-stabilized, photochemically prepared silver nanoparticles. Biomaterials. 2012;33(19):4947–56.

53. Alarcon EI, Udekwu KI, Noel CW, Gagnon LBP, Taylor PK, Vulesevic B, Simpson MJ, Gkotzis S, Islam MM, Lee CJ and others. Safety and efficacy of composite collagen-silver nanoparticle hydrogels as tissue engineering scaffolds. Nanoscale. 2015;7(44):18789–98.
54. Poblete H, Agarwal A, Thomas SS, Bohne C, Ravichandran R, Phospase J, Comer J, Alarcon EI. New insights into peptide-silver nanoparticle interaction: deciphering the role of cysteine and lysine in the peptide sequence. Langmuir. 2016;32(1):265–73.
55. Pokhrel LR, Dubey B, Scheuerman PR. Impacts of select organic ligands on the colloidal stability, dissolution dynamics, and toxicity of silver nanoparticles. Environ Sci Technol. 2013;47(22):12877–85.
56. Seitz F, Rosenfeldt RR, Storm K, Metreveli G, Schaumann GE, Schulz R, Bundschuh M. Effects of silver nanoparticle properties, media pH and dissolved organic matter on toxicity to daphnia magna. Ecotoxicol Environ Saf. 2015;111:263–70.
57. Sharma VK, Siskova KM, Zboril R, Gardea-Torresdey JL. Organic-coated silver nanoparticles in biological and environmental conditions: fate, stability and toxicity. Adv Colloid Interface Sci. 2014;204:15–34.
58. Vignoni M, Weerasekera HDA, Simpson MJ, Phopase J, Mah TF, Griffith M, Alarcon EI, Scaiano JC. LL37 peptide@silver nanoparticles: combining the best of the two worlds for skin infection control. Nanoscale. 2014;6(11):5725–8.
59. Carlson C, Hussain SM, Schrand AM, Braydich-Stolle LK, Hess KL, Jones RL, Schlager JJ. Unique cellular interaction of silver nanoparticles: size-dependent generation of reactive oxygen species. J Phys Chem B. 2008;112(43):13608–19.
60. Kwan KHL, Liu XL, To MKT, Yeung KWK, Ho CM, Wong KKY. Modulation of collagen alignment by silver nanoparticles results in better mechanical properties in wound healing. Nanomedicine. 2011;7(4):497–504.
61. Mishra M, Kumar H, Tripathi K. Diabetic delayed wound healing and the role of silver nanoparticles. Dig J Nanomater Bios. 2008;3(2):49–54.
62. Alarcon EI, Vulesevic B, Argawal A, Ross A, Bejjani P, Podrebarac J, Ravichandran R, Phopase J, Suuronen EJ, Griffith M. Coloured cornea replacements with anti-infective properties: expanding the safe use of silver nanoparticles in regenerative medicine. Nanoscale. 2016;8(12):6484–9.
63. Dong RH, Jia YX, Qin CC, Zhan L, Yan X, Cui L, Zhou Y, Jiang XY, Long YZ. In situ deposition of a personalized nanofibrous dressing via a handy electrospinning device for skin wound care. Nanoscale. 2016;8(6):3482–8.
64. Ziv-Polat O, Topaz M, Brosh T, Margel S. Enhancement of incisional wound healing by thrombin conjugated iron oxide nanoparticles. Biomaterials. 2010;31(4):741–7.
65. Trickler WJ, Lantz SM, Schrand AM, Robinson BL, Newport GD, Schlager JJ, Paule MG, Slikker W, Biris AS, Hussain SM and others. Effects of copper nanoparticles on rat cerebral microvessel endothelial cells. Nanomedicine. 2012;7(6):835–46.
66. Chen WY, Chang HY, Lu JK, Huang YC, Harroun SG, Tseng YT, Li YJ, Huang CC, Chang HT. Self-assembly of antimicrobial peptides on gold nanodots: against multidrug-resistant bacteria and wound-healing application. Adv Funct Mater. 2015;25(46):7189–99.
67. Chigurupati S, Mughal MR, Okun E, Das S, Kumar A, McCaffery M, Seal S, Mattson MP. Effects of cerium oxide nanoparticles on the growth of keratinocytes, fibroblasts and vascular endothelial cells in cutaneous wound healing. Biomaterials. 2013;34(9):2194–201.
68. Das S, Baker AB. Biomaterials and nanotherapeutics for enhancing skin wound healing. Front Bioeng Biotechnol. 2016;4.
69. Unnithan AR, Sasikala ARK, Sathishkumar Y, Lee YS, Park CH, Kim CS. Nanoceria doped electrospun antibacterial composite mats for potential biomedical applications. Cer Int. 2014;40(8):12003–12.
70. Elsabahy M, Wooley KL. Design of polymeric nanoparticles for biomedical delivery applications. Chem Soc Rev. 2012;41(7):2545–61.
71. Metcalfe AD, Ferguson MWJ. Tissue engineering of replacement skin: the crossroads of biomaterials, wound healing, embryonic development, stem cells and regeneration. J R Soc Interface. 2007;4(14):413–37.

72. Mogosanu GD, Grumezescu AM. Natural and synthetic polymers for wounds and burns dressing. Int J Pharm. 2014;463(2):127–36.
73. Chereddy KK, Coco R, Memvanga PB, Ucakar B, des Rieux A, Vandermeulen G, Preat V. Combined effect of PLGA and curcumin on wound healing activity. J Control Release. 2013;171(2):208–15.
74. Mun SH, Joung DK, Kim YS, Kang OH, Kim SB, Seo YS, Kim YC, Lee DS, Shin DW, Kweon KT and others. Synergistic antibacterial effect of curcumin against methicillin-resistant Staphylococcus aureus. Phytomedicine. 2013;20(8–9):714–18.
75. Durgaprasad S, Reetesh R, Hareesh K, Rajput R. Effect of a topical curcumin preparation (BIOCURCUMAX) on burn wound healing in rats. J Pharm Biomed Sci 2011;8(08).
76. Chu YJ, Yu DM, Wang PH, Xu J, Li DQ, Ding M. Nanotechnology promotes the full-thickness diabetic wound healing effect of recombinant human epidermal growth factor in diabetic rats. Wound Repair Regen. 2010;18(5):499–505.
77. Nafee N, Youssef A, El-Gowelli H, Asem H, Kandil S. Antibiotic-free nanotherapeutics: hypericin nanoparticles thereof for improved in vitro and in vivo antimicrobial photodynamic therapy and wound healing. Int J Pharm. 2013;454(1):249–58.
78. Angelova N, Yordanov G. Nanoparticles of poly(styrene-co-maleic acid) as colloidal carriers for the anticancer drug epirubicin. Colloids Surf A. 2014;452:73–81.
79. Keeney M, Ong SG, Padilla A, Yao ZY, Goodman S, Wu JC, Yang F. Development of poly(beta-amino ester)-based biodegradable nanoparticles for nonviral delivery of minicircle DNA. ACS Nano. 2013;7(8):7241–50.
80. Park HJ, Lee J, Kim MJ, Kang TJ, Jeong Y, Um SH, Cho SW. Sonic hedgehog intradermal gene therapy using a biodegradable poly (beta-amino esters) nanoparticle to enhance wound healing. Biomaterials. 2012;33(35):9148–56.
81. Archana D, Dutta J, Dutta PK. Evaluation of chitosan nano dressing for wound healing: characterization, in vitro and in vivo studies. Int J Biol Macromol. 2013;57:193–203.
82. Gao WI, Lai JCK, Leung SW. Functional enhancement of chitosan and nanoparticles in cell culture, tissue engineering, and pharmaceutical applications. Front Physiol. 2012;3.
83. Rnjak-Kovacina J, Weiss AS. The role of elastin in wound healing and dermal substitute design. In: Dermal replacements in general, burn, and plastic surgery. New York: Springer; 2013. p. 57–66.
84. Koria P, Yagi H, Kitagawa Y, Megeed Z, Nahmias Y, Sheridan R, Yarmush ML. Self-assembling elastin-like peptides growth factor chimeric nanoparticles for the treatment of chronic wounds. PNAS. 2011;108(3):1034–9.
85. Kwon MJ, An S, Choi S, Nam K, Jung HS, Yoon CS, Ko JH, Jun HJ, Kim TK, Jung SJ and others. Effective healing of diabetic skin wounds by using nonviral gene therapy based on minicircle vascular endothelial growth factor DNA and a cationic dendrimer. J Gene Med. 2012;14(4):272–8.
86. Zarrintaj P, Moghaddam AS, Manouchehri S, Atoufi Z, Amiri A, Amirkhani MA, Nilforoushzadeh MA, Saeb MR, Hamblin MR, Mozafari M. Can regenerative medicine and nanotechnology combine to heal wounds? The search for the ideal wound dressing. Nanomedicine. 2017;12(19):2403–22.
87. Bhagat V, Becker ML. Degradable adhesives for surgery and tissue engineering. Biomacromol. 2017;18(10):3009–39.
88. Pupkaite J, Ahumada M, McLaughlin S, Temkit M, Alaziz S, Seymour R, Ruel M, Kochevar I, Griffith M, Suuronen EJ and others. Collagen-based photoactive agent for tissue bonding. ACS Appl Mater Interfaces. 2017;9(11):9265–70.
89. Xu N, Yao M, Farinelli W, Hajjarian Z, Wang Y, Redmond RW, Kochevar IE. Light-activated sealing of skin wounds. Lasers Surg Med. 2015;47(1):17–29.
90. Zhao X, Sun X, Yildirimer L, Lang Q, Lin ZYW, Zheng R, Zhang Y, Cui W, Annabi N, Khademhosseini A. Cell infiltrative hydrogel fibrous scaffolds for accelerated wound healing. Acta Biomater. 2017;49:66–77.
91. Nizamoglu S, Gather MC, Humar M, Choi M, Kim S, Kim KS, Hahn SK, Scarcelli G, Randolph M, Redmond RW and others. Bioabsorbable polymer optical waveguides for deep-tissue photomedicine. Nat Comm. 2016;7.

92. Han S, Hwang BW, Jeon EY, Jung D, Lee GH, Keum DH, Kim KS, Yun SH, Cha HJ, Hahn SK. Upconversion nanoparticles/hyaluronate-rose bengal conjugate complex for noninvasive photochemical tissue bonding. ACS Nano. 2017;11(10):9979–88.
93. Singh M, Nuutila K, Kruse C, Robson MC, Caterson E, Eriksson E. Challenging the conventional therapy: emerging skin graft techniques for wound healing. Plast Reconstr Surg. 2015;136(4):524E–30E.
94. Tam J, Wang Y, Farinelli WA, Jimenez-Lozano J, Franco W, Sakamoto FH, Cheung EJ, Purschke M, Doukas AG, Anderson RR. Fractional skin harvesting: autologous skin grafting without donor-site morbidity. Plast Reconstr Surg Glob Open. 2013;1(6):e47.
95. Tam J, Wang Y, Vuong LN, Fisher JM, Farinellil WA, Anderson RR. Reconstitution of full-thickness skin by microcolumn grafting. J Tissue Eng Regen Med. 2017;11(10):2796–805.
96. Franco W, Jimenez-Lozano JN, Tam J, Purschke M, Wang Y, Sakamoto FH, Farinelli WA, Doukas AG, Anderson RR. Fractional skin harvesting: device operational principles and deployment evaluation. J Med Dev. 2014;8(4).
97. Tam J, Farinelli W, Franco W, Anderson RR. Apparatus for harvesting tissue microcolumns. J Vis Exp. 2018;(140).

Chapter 7
Nano-engineering Nanoparticles for Clinical Use in the Central Nervous System: Clinically Applicable Nanoparticles and Their Potential Uses in the Diagnosis and Treatment of CNS Aliments

Suzan Chen, Angela Auriat, Anna Koudrina, Maria DeRosa, Xudong Cao and Eve C. Tsai

Abstract Nano-engineering materials-based diagnosis and treatment of central nervous systems (CNS) ailments has significantly advanced with our deepened knowledge of the pathophysiology of the blood–brain barrier. Unlike other nanoparticle-based tissue engineering strategies, the use of nanoparticles in the CNS must be specifically engineered to circumvent or penetrate the blood–brain barrier, which selectively inhibits drugs and nanoparticles from infiltrating. Current research in the field of CNS nanoparticles has future applications in the fields of diagnostic imaging, drug delivery, specific drug targeting, and tissue regeneration. This chapter highlights some of the nano-engineering of these promising nanoparticle-based biomaterials and their applications in the diagnosis and treatment of brain and spinal cord disease.

S. Chen · A. Auriat · E. C. Tsai (✉)
Ottawa Hospital Research Institute, Loeb Building,
725 Parkdale Ave, Ottawa, ON K1Y 4E9, Canada
e-mail: etsai@toh.ca

A. Koudrina · M. DeRosa
Department of Chemistry and Institute of Biochemistry, Carleton University,
225 Steacie Building 1125 Colonel by Drive, Ottawa, ON K1S 5B6, Canada

X. Cao
Faculty of Engineering, University of Ottawa, Colonel by Hall 161 Louis Pasteur Private,
Room CBY A-337, Ottawa, ON K1N 6N5, Canada

E. C. Tsai
Division of Neurosurgery, Suruchi Bhargava Chair in Spinal Cord and Brain Regeneration
Research, University of Ottawa, The Ottawa Hospital—Civic Campus C2, 1053 Carling Ave,
Ottawa, ON K1Y 4E9, Canada

© Springer Nature Switzerland AG 2019
E. I. Alarcon and M. Ahumada (eds.), *Nanoengineering Materials for Biomedical Uses*,
https://doi.org/10.1007/978-3-030-31261-9_7

7.1   Introduction

While significant advances in our understanding of the pathophysiology of the central nervous system (CNS) have occurred, our diagnosis and treatment strategies of CNS ailments remain limited. Diagnosis is still challenging as small lesions in an eloquent region of the brain, or spinal cord can lead to significant deficits and large lesions that do not affect eloquent regions of the brain may not result in any clinically noticeable deficits or symptoms. Our ability to visualize these small lesions remains limited with current clinical imaging technology. While we have also made significant strides in our ability to offer therapeutic strategies, many of these agents cannot be applied systemically as they are inactivated with systemic administration, cannot cross the blood–brain barrier (BBB) [1, 2], are rapidly degraded in the circulatory system [3], or cause unacceptable systemic side effects [4]. Accordingly, CNS-targeted nanoparticles may be the answer to these barriers, improving the diagnosis and treatment of CNS disorders.

7.1.1   The Blood–Brain Barrier: A Physiological Barrier to Treatment

The blood–brain barrier (Fig. 7.1) is a physical and chemical barrier which protects the CNS and offers homeostasis. Formed from capillary endothelium, astrocytes, pericytes, and extracellular matrix, the BBB allows for highly controlled diffusion of essential compounds from the bloodstream into the CNS, such as water, lipid-soluble molecules, and gasses while restricting the entry of pathogenic organisms and other potentially harmful substances, including hydrophilic molecules >1 kDa and exogenous molecules. These functions are largely dependent on the presence of a variety of components, such as tight junctions which are highly selective in order

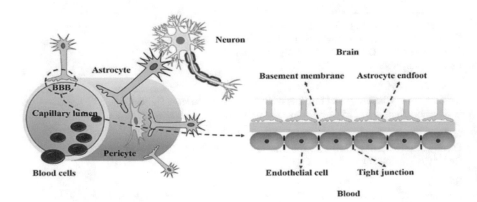

Fig. 7.1 Schematic of blood–brain barrier Adapted with permission from Zhou et al. [7]

to limit the passage of solutes, ATP-binding cassette transporters (ABC transporters) which actively pump xenobiotics out of the CNS [5], and endothelial cytochromes and enzymes, which degrade small-sized molecules [6].

Circumventing the BBB presents a standing challenge for the diagnosis and treatment of many CNS-related ailments. Over 95% of pharmaceuticals which could diagnose or treat CNS-related diseases cannot cross the BBB, and while many of these treatments have been successful in animal models of disease, they cannot be translated for human use due to failure of drug delivery [8, 9]. This is particularly challenging when repeated administration is required for sustaining a therapeutic dose [10].

Current delivery of CNS-specific therapeutics across the BBB can be categorized as physical, chemical, or biological and can be used for CNS-specific diseases. Physical delivery can be achieved by using transcranial drug delivery methods to pass drugs directly into the brain, physically trespassing the BBB. These methods involve neurosurgical procedures which can be accomplished via intracerebral implantation (Fig. 7.2a), intracerebral infusion (Fig. 7.2b), and convection-enhanced diffusion (Fig. 7.2c). Intracerebral infusion and intracerebral implantation are both able to bypass the BBB with physical penetration. However, studies have indicated that as the drug penetrates the brain tissue, the concentration of bioavailable drugs decreases significantly with each millimeter of tissue away from the site of injection or implant. By their definition, these methods are invasive and require the patient to undergo a surgical procedure [11].

Chemical-based drug delivery is based on causing a break in the BBB which allows drugs to penetrate through the tight junctions. This BBB disruption can occur by utilizing:

- Osmotic disruptors such as bradykinins
- Use of high-intensity ultrasound
- Introducing osmotic pressure or microbubbles.

These methods, while novel, are potentially harmful to patients as they may permanently damage the integrity of the BBB, allowing not only the targeted pharmaceutical but other potentially harmful substances to cross [7]. Biological-based drug delivery utilizes the targeting of a drug to a ligand which can bind to an endocytic receptor or lipophilic analog to improve BBB penetration or diffusion [12]. However, studies have shown that all these methods suffer from a range of limitations, including but not limited to lack of specificity, resulting in adverse effects on healthy tissues, and loss of bioavailable concentration, necessitating repeated exposure, which results in further adverse effects [10].

7.1.2 Nanoparticle-Based Delivery

Nanoparticles are small structures on the nanoscale (1–100 nm, see Chap. 1). They can be utilized to carry and deliver therapeutic agents in complex biological systems

(a)

(b)

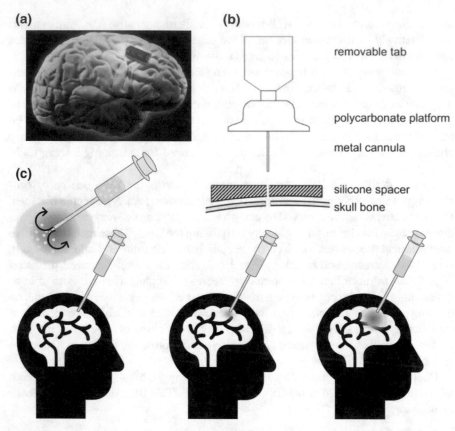

removable tab

polycarbonate platform

metal cannula

silicone spacer

skull bone

(c)

Fig. 7.2 Physical delivery methods through the BBB. **a** Intracerebral implantation: involves place-ment of the target drug or delivery system directly on the brain's surface. **b** Cannula for intracerebral infusion. This cannula would require surgical insertion through the skull bone to enter the brain. **c** Convection-enhanced diffusion: Using a pump, target drug is infused through a cannula inserted into the target with the addition of continuous positive pressure

as therapeutic nano-carriers. These carriers have been fabricated from many different materials and are highly customizable in regard to size, charge, and molecular prop-erties [10]. Nanoparticles are the ideal molecules for CNS-specific drug delivery. Recent studies have focused on the creation of biodegradable nanoparticles which are able to increase therapeutic bioavailability, retention, and solubility of diagnostic and therapeutic agents. These agents can be encapsulated by or embedded within the surface of nanoparticles, which protects the drug from degradation and allows for extended release. Payload-carrying nanoparticles can be further modified with receptor-targeting ligands, which results in targeting of the whole system to specific tissues. Due to their versatility, nanoparticle-based delivery systems are currently being studied for CNS-related clinical applications in cancers, acquired immunode-ficiency syndrome (AIDS), as well as non-CNS-related clinical applications such as non-CNS cancers, diabetes, malaria, and tuberculosis [10].

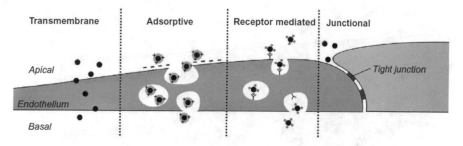

Fig. 7.3 Routes for movement of nanoparticles across the brain endothelium reproduced with permission from Male et al. [10]

Table 7.1 Summary of active and passive transport of nanoparticle-based CNS drug delivery

	Active transport			Passive transport
Mechanism	Receptor-mediated	Adsorptive	Carrier-mediated transport	Transmembrane
Targets	Ligands for receptors	Cationic nanoparticles	Ligands for transport proteins	Small lipophilic molecules < 400-500 Da
Examples	PBCA PLGA/PLA Transferrin-conjugated nanoparticles Liposomes	Liposomes	PLGA/PLA carbon quantum dots	Gold nanoparticles Sliver nanoparticles

Nanoparticle-based delivery of pharmaceuticals to the CNS offers many benefits when compared to traditional physical, chemical, and biological methods of cross-BBB delivery. Nanoparticles can offer a noninvasive, low-cost, and highly controlled method to load and release these targeted drugs across the BBB [10].

Transport across the BBB can both be passive or active. Passive transport is energy-independent and utilizes the process of passive diffusion. This is particularly useful for tumor cells due to the enhanced permeability and retention effect. In contrast, active transport requires the use of carriers at the expense of adenosine triphosphate (ATP) via a receptor and adsorption-mediated routes (Fig. 7.3). These differences, with examples of nanoparticles which will be further discussed below, are highlighted in Table 7.1.

7.2 Polymeric Nanoparticles

Polymeric nanoparticles (Fig. 7.4) are composed of natural or synthetic polymers and can be fabricated into nanoparticles that are suitable for CNS delivery.

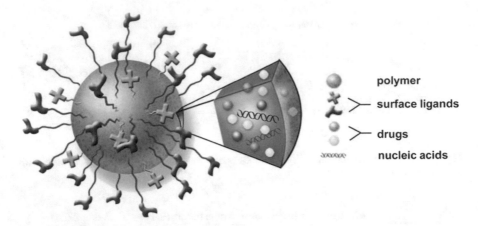

Fig. 7.4 Schematic of a functionalized polymeric nanoparticle. Reproduced with permission from Patel et al. [13]

These nanoparticles have been based on polysaccharides, proteins, amino acids, poly(ethylenimines), poly(alkylcyanoacrylates), poly(-methylidene malonates), and polyesters [13]. There are currently two polymeric-based nanoparticle drug delivery systems which are Food and Drug Administration (FDA) approved for clinical use:

1. Abdoscan®: An iron oxide and dextran-based nanoparticle used for diagnostic imaging of the liver and spleen [14].
2. Abraxane®: An albumin-based nanoparticle loaded with paclitaxel used in the treatment for breast cancer [15].

The basic structure of a polymeric nanoparticle is illustrated in Fig. 7.4. The drug, which could be a synthetic molecule, a peptide, or a nucleic acid in nature, is embedded within the walls of the polymer, while the surface is modified with a ligand, which is intended to deliver the nanoparticle to the area of interest. In the context of treatments of CNS ailments, the ligand would specifically target a component within the BBB. Systemic delivery of these nanoparticles across the BBB is possible because their surface can be so readily modified, relying on ligand surface-modification for receptor-mediated transcytosis or surface charge-modification for adsorptive-mediated transcytosis [13]. Unfortunately, these favorable characteristics are still restricted by potential toxicity, including both chemical toxicity determined by the purity and concentration of these nanoparticles, along with nano-toxicity resulting from particle size, shape, surface, and charge [16].

7.2.1 Poly(Butyl cyanoacrylate) Nanoparticles

First synthesized in 1995, poly(butyl cyanoacrylate) (PBCA, see Fig. 7.5) nanoparticles were the first polymer-based nanoparticle system studied to deliver drugs to

Fig. 7.5 Chemical structure
of poly(butyl cyanoacrylate)

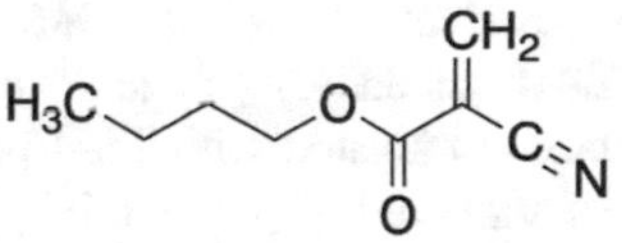

the CNS through the BBB [17]. PBCA nanoparticles were loaded with dalargin, an opioid peptide Leu-enkephalin with analgesic properties, coated with polysorbate 80, and delivered intravenously and orally. The results showed that these nanoparticles are able to achieve analgesia in the CNS of a mouse model [18]. Follow-up studies, using radiolabeled particles, demonstrated that in the absence of polysorbate-80 coating, there was a significant decrease in the number of PBCA nanoparticles that crossed the BBB [19]. PBCA nanoparticles coated by polysorbate-80 can also be loaded with fluorophores, antibodies, or magnetic contrast agents, with possible clinical applications in diagnostic imaging of the CNS [20]. However, non-specific permeabilization of the BBB, probably related to the toxicity of the carrier, caused mortality in mice, questioning the clinical application of this nanoparticle [16].

7.2.2 Poly(Lactic-Co-glycolic Acid)/Poly(Lactic Acid) Nanoparticles

Poly(lactic-*co*-glycolic acid)/poly(lactic acid) (PLGA/PGA)-based nanoparticles, like PBCA, have been in development since the 1990s, due to their promising biocompatibility and biodegradability [21]. Their chemical structures can be seen in Fig. 7.6. PLGA/PLA can be hydrolyzed by the body into lactic acid or glycolic acid, respectively, and metabolized by the Kreb's cycle [22]. PLGA has many current clinical uses such as suture material. PLGA/PGA nanoparticles of different sizes, size distribution, morphology, and ζ potential can be synthesized by controlling the parameters specific to the synthesis method employed [23]. PLGA/PLA nanoparticles can also undergo surface-modification with surfactants or polymers, or covalent conjugation with targeting ligands which can improve BBB penetration. This high degree of customization makes PLGA/PGA nanoparticles applicable to a wide range of treatments within the CNS [22].

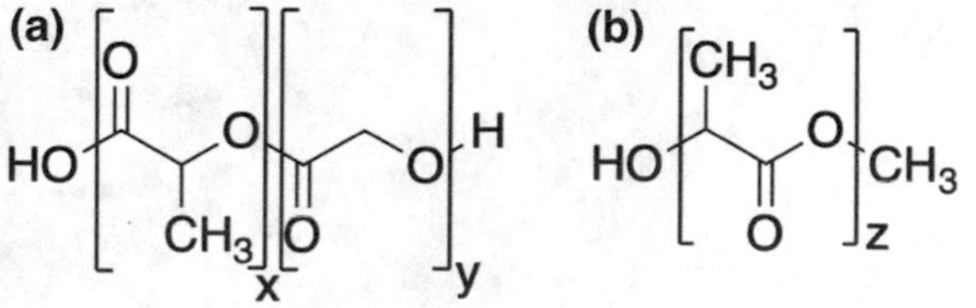

Fig. 7.6 **a** Chemical structure of poly(lactic-co-glycolic acid); x and y denote the number of unit repeats. **b** Chemical structure of poly(lactic acid): z denotes the number of unit repeats

Recent research has shown that PLGA nanoparticles conjugated with cyclopeptides can deliver zinc ions via endocytosis [24]. Curcumin-loaded PLGA nanoparticles conjugated with Tet-1 peptide have shown promising ability to cross the BBB in vitro and can potentially be applied for treating Alzheimer's dementia [25]. Paclitaxel-loaded PLGA nanoparticles have recently been produced by Lei et al. demonstrating excellent reproducibility and uptake in the mouse model [26].

7.2.2.1 PBCA/PLGA/PLA Nanoparticles as Drug Carriers for Alzheimer's Disease

Alzheimer's disease (AD) is a common and devastating type of dementia which is characterized by learning and memory impairment. There are about 4 million people living with dementia worldwide, and AD accounts for an estimated 60–80% of these cases [27].

Curcumin is a diarylheptanoid, which is a plant-based compound with biological activity against β-amyloid (Aβ) aggregate, shown in Fig. 7.7, which is the main component of neuronal amyloid plaques causing AD [28]. Unfortunately, curcumin has poor aqueous solubility and stability and is prone to oxidation and photodegradation. These properties make it a poor candidate for oral, systematic administration. PBCA nanoparticles embedded with curcumin have shown to improve the drug's photostability in a SH-SY5Y cell culture model [29]. The most promising of the nanoparticle curcumin carriers had sustained stability, with long-term (6 months) storage, while remaining bioactive [30]. In vitro experiments demonstrated that this system has the ability to destroy amyloid plaques [25], while in a mouse model, it was able to improve cue memory and lower amyloid plaque activity [31].

Fig. 7.7 Amyloid plaque, scale bar represents 50 μm Adapted from Mathur et al. [32]

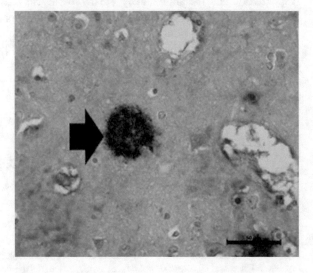

7.2.3 Carbon Quantum Dots

Carbon quantum dots have been discovered in 2004 and are the newest member of the polymeric nanoparticle family discussed in this chapter [33]. Carbon quantum dots are a promising type of nanoparticle due to their low toxicity, small relative size, polymeric core, and the available surface functional groups [33]. These surface functional groups are particularly useful for conjugation with therapeutics for the purposes of drug delivery across the BBB [34, 35]. Quantum dots also have excellent photoluminescence, allowing for real-time tracking of BBB penetration shown in animal models [36].

Carbon quantum dots are nano-fabricated by either "top-down" or "bottom-up" synthesis. The top-down approach involves oxidizing double bonds from macromolecular starting materials such as raw carbon powder [37], while the bottom-up approach involves building carbon quantum dots from small monomeric units such as citric acid and amines using covalent hydrogen bonds [37]. Top-down carbon quantum dots are synthesized using raw carbon powder, which are then conjugated to transferrin and doxorubicin, making them a promising treatment agent for pediatric CNS neoplasms. Doxorubicin is a chemotherapeutic agent which is effective against many different cancers including breast, bladder, lymphoma, and sarcoma [38]. However, doxorubicin, like many chemotherapeutics, has side effects such as hair loss and bone marrow suppression, and may even cause congestive heart failure in the pediatric population [38]. Carbon quantum dots conjugated with doxorubicin and transferrin were produced in a step-wise process. First, raw carbon dots were synthesized and purified, which was followed by conjugation of transferrin and subsequent conjugation of doxorubicin, utilizing the available surface functional groups [38]. Transferrin was chosen due to the overexpression of the transferrin receptor on the BBB and tumor cells of interest. Successful uptake of these nanoparticles by the pediatric tumor cell line, SJGMB, has been seen in vitro when compared to doxorubicin alone [38].

Bottom-up carbon quantum dots synthesized using L-aspartic acid and D-glucose via pyrolysis have the ability to act as a targeted imaging and diagnostic agent for gliomas when injected intravenously in a mouse model [37]. This is due to the carbon quantum dot's excitation-dependent photoluminescence behavior [37]. It is hypothesized that these carbon quantum dots are transported across the BBB with the assistance of transport proteins, such as GLUT1 and ACST2, using the glucose and aspartic acid ligands [39]. This hypothesis can be supported by the known abundance of these transporters, specifically found on the surface of the BBB and gliomas [40].

7.2.4 Liposomes

Liposomes are spherical vesicles that consist of one or more lipid bilayers, which form an internal aqueous compartment. Liposomes have relatively impermeable lipophilic outer shells and are often composed of phospholipids. They have been among the most investigated structures for drug delivery in recent history because of their excellent biocompatibility, biodegradability, low toxicity properties and ability to incorporate both hydrophilic and hydrophobic drugs. Liposome-based drug delivery systems are based on the aqueous core, which provides an environment in which therapeutic drugs can be sequestered and have been widely used to improve drug efficacy or to eliminate drug-related toxicity [41].

Adsorptive-mediated transcytosis of liposome-based drug delivery systems utilizes the negatively charged property of endothelial cells. By creating positively charged or "cationized" liposomes, these liposomes are able to target and bind to the BBB, mediating the incorporation of their contents across the BBB. Studies have shown that liposomes conjugated with cationized bovine serum albumin are able to adsorb onto the BBB in in vivo rat models, while liposomes conjugated with naive bovine serum albumin did not achieve similar levels of adsorption [42].

7.2.4.1 Targeted Delivery of Liposomes to CNS

The process of delivery of liposomes to a target is generally complex and non-specific. An approach referred to as "Molecular Trojan Horse" has proven to be effective in delivering molecules through the blood–brain barrier to the brain by exploiting receptor-mediated transcytosis. This method consists of surface-modifying the exterior liposomes to include ligands specific to receptors present on the target of interest. Different ligands have been researched and include peptides, monoclonal antibodies and aptamers. As mentioned above, one common receptor that is present on the surface of the brain capillary endothelium is transferrin receptor (TfR). TfR is responsible for the transport of holo-transferrin from blood to the brain, at the same time as mediating reverse transcytosis of apo-transferrin from the brain to blood [43]. Recognition elements, such as transferrin or antitransferrin receptor antibody, and a transferrin receptor-specific aptamer have been used to effectively and specifically deliver the liposomal payload, such as anticancer drugs, in order to improve the specificity of cellular uptake while lowering overall toxicity [44].

7.2.4.2 Liposomes for Treatment of CNS Infection

One promising CNS application of liposome-mediated drug delivery is based on increasing the bioavailability of antiretroviral pharmaceuticals for the treatment of CNS infection. Generally, loading efficiency of anti-HIV drugs has been reported to be low, while the portion that is successfully loaded suffers from a high amount of leakage. In a study performed by Li and colleagues, liposomes were filled with a zidovudine (AZT) prodrug and zidovudine myristate (AZT-M), in order to avoid leakage, and were subsequently injected intravenously into the rat model. Although higher concentrations of AZT were found in the brain post-AZT-M liposomal injection when compared to the free AZT injection, the greatest increase of the drug was found in the liver and spleen. This is important for this type of treatment, because drug accumulation in the reticuloendothelial system is favorable, as it improves the therapeutic index of antiretroviral pharmaceuticals [45].

7.2.4.3 Liposomes for Treatment of CNS Neoplasia

One of the shortcomings of many anticancer drugs is that they suffer from a lack of specificity, which ultimately results in cytotoxicity towards not only the tumor cells but also towards the healthy cells. In order to reduce the severe systemic toxic side effects, these anticancer drugs can be incorporated into a targeted payload-carrying system, which will ultimately reduce non-specific toxicity and improve the efficacy of the treatment [46]. Transferrin can once again be used as a target in CNS neoplasia treatments, as it is highly expressed on the surface of tumor cells. For example, antitransferrin receptor antibody RI7217 can be conjugated to the surface of liposomes, which would not only allow for targeted delivery of the payload-carrying liposomes to BBB, resulting in their subsequent fusion and release of the payload, but also for targeting transferrin presented on the surface of tumor cells [47]. One example of the use of liposomal formulation in the treatment of neoplastic meningitis was examined by Dominguez et al., when the liposomes were loaded with cytarabine, a cell-specific antimetabolite that is used to kill tumor cells, while they are in their S-phase of the cell cycle. Cytarabine was encapsulated in the aqueous compartments of the liposomal wall, made up of phospholipids, triglycerides and cholesterol. Therapeutic efficacy of cytarabine is largely dependent on the presence of a large dose of the drug during a specific phase of cell development, which means that the concentration of the drug and the length of treatment must be carefully tailored. Liposomal cytarabine showed sustained release of the drug, helping to maintain the proper concentration, which spanned the cell cycle, allowing specific S-phase targeting [48].

7.2.4.4 Liposomes for Treatment of Ischemic Stroke

Stroke has a worldwide incidence of 15 million new cases each year and a mortality rate of approximately 30% [49]. Ischemic stroke accounts for approximately 80% of all stroke events. Liposome-based therapeutic strategies that are discussed below differ from the ones above as they are specifically targeted to the vasculature of the brain and do not require the penetration of the BBB.

Liposomes prepared from dipalmitoylphosphatidylcholine (DPPC)/dioleoylphosphatidylcholine (DOPC)/cholesterol (Chol) have shown promise in the acute neuroprotection during ischemic stroke. These liposomes allow for a controlled release of nitric oxide (NO), a regulator of cerebral artery tone and a neuroprotector. NO is usually applied to a patient systematically and can be scavenged by hemoglobin. These liposomes are able to deliver NO specifically to the area of injury and achieve a high local concentration, allowing for increased benefit and neuroprotection [50].

Another group, led by Hwang, was able to create a liposome composed of a mixture of phosphatidylcholine (PC), Chol and phosphatidylethanolamine. This liposome can be used to deliver angiogenic peptides derived from the vascular endothelial growth factor to an ischemic brain in a rat model [51]. This liposome can be administered intra-arterially and is able to target the ischemic hemisphere of the brain. The application of these liposomes resulted in the attenuation of the perfusion defect and increased expression of a gene involved in angiogenesis (angiopoietin-2), with the consequent increase of glucose consumption and vascular density, without promoting inflammation [51].

7.2.4.5 Liposomes for Treatment of Cognitive Deficits

Another payload/targeting liposome system was developed by McConnell and colleagues, where a transferrin-targeting aptamer (TRA) was used to surface-modify a liposome, which was loaded with a dopamine aptamer (DAL). This dual-aptamer system (DAL-TRAM), with one aptamer acting as the mediator and the other as the payload, allowed for specific targeting and take-up of the liposome through the BBB and subsequent safe delivery of the nucleic acid agent. Other advantages of this system include the ability to modify the composition of liposome walls to include fluorescent elements that can be used in real-time imaging of the distribution, and the delivery of a high local concentration of the payload at the target of interest. This particular study confirmed the ability of an acute systemic administration of DAL-TRAM to attenuate hyperlocomotion in cocaine-treated mice. This was achieved by the TRA-driven specific delivery of a dopamine aptamer to the brain [44].

7.2.4.6 Nanoparticle for Improved Neural Axis Imaging

Magnetic resonance imaging (MRI) has revolutionized the diagnosis and management of CNS disease, and contrast agents (CA) have been heavily used over the past several decades to enhance the diagnostic value of the obtained images. Improving the efficacy of contrast agents can be facilitated by allowing both the optimization of the magnetic properties of the CA and the optimization of the pharmacokinetics and distribution of the CA in the patient. Contrast agents consisting of DNA aptamer-gadolinium(III) conjugates have been shown to provide a single system in which these factors can be addressed simultaneously. We have shown that a 15mer thrombin aptamer could be conjugated to diethylenetriaminepentaacetic (DTPA) dianhydride to form a monoamide derivative of the linear open-chain chelate present in the commonly used contrast agent Magnevist($^®$) [52]. The stability of the conjugated DNA aptamer-DTPA-Gd(III) chelate in a transmetallation study using Zn(II) was found to be similar to that reported for DTPA-Gd(III). Relaxivity enhancements of 35 ± 4 and 20 ± 1% were observed in the presence of thrombin compared to a control protein at fields of 9.4 and 1.5 T, respectively.

Multifunctional nanoparticles have also been developed towards applications in noninvasive magnetic resonance imaging and axonal tracing. We have developed multifunctional nano-biomaterial by deliberately combining functions of superparamagnetism, fluorescence and axonal tracing into one material [53]. Superparamagnetic iron oxide nanoparticles were first synthesized and coated with a silica layer to prevent emission quenching through core–dye interactions; a fluorescent molecule, fluorescein isothiocyanate, was doped inside second layer of silica shell to improve photostability and to enable further thiol functionalization. Subsequently, biotinylated dextran amine, a sensitive axonal tracing reagent, was immobilized on the thiol-functionalized nanoparticle surfaces. The resulting nanoparticles were then characterized by transmission electron microscopy, dynamic light scattering, X-ray diffraction, X-ray photoelectron spectroscopy, UV–Vis spectroscopy, magnetic resonance imaging and fluorescence confocal microscopy. In vitro cell experiments using both undifferentiated and differentiated Neuro-2a cells showed that the cells were able to take up the nanoparticles intracellularly and that the nanoparticles showed good biocompatibility. This new material demonstrated promising performances for both optical and magnetic resonance imaging modalities, suggesting its promising potential in applications such as in noninvasive imaging, particularly with respect to neuronal tracing.

7.3 Inorganic Nanoparticles

7.3.1 Gold Nanoparticles

Gold nanoparticles, shown in Fig. 7.8, (<10 nm) have a long-standing history in human use, dating to medieval times when they were used to create decorative glass. Gold nanoparticles possess a "surface plasmon resonance" phenomenon, which is a non-radiative electromagnetic surface wave that propagates in a direction parallel to the negative permittivity/dielectric material interface. This allows for the measuring of the adsorption of the nanoparticle. Ligands can be attached to the nanoparticles during synthesis via a thiol bond or through an exchange reaction after synthesis with thiolated ligands. Gold nanoparticles are also easy to prepare and can undergo versatile surface-modifications while being highly biocompatible. Gold nanoparticles have a number of advantages as both an imaging reagent and a therapeutic delivery system in the CNS [10].

Fig. 7.8 Representative
TEM image of a gold
nanoparticle

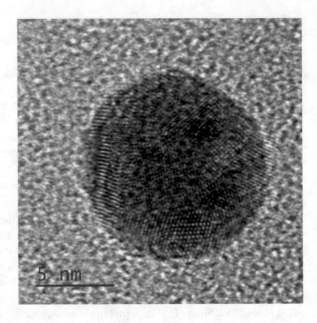

5 nm

7.3.1.1 Gold Nanoparticles for Diagnoses of CNS Tumors

Currently used as X-ray contrast agents, gold nanoparticles are advantageous over iodine as they are less nephrotoxic and produce higher contrast enhancement [54]. Gold nanoparticles can be used to specifically target tumors and are used in the imaging of kidney carcinomas. However, gold nanoparticles, to date, have shown a low level of accumulation in the brain. Therefore, more is required to realize the potential of gold nanoparticles in relation to diagnostics, especially with respect to CNS tumors [10].

7.3.1.2 Gold Nanoparticles for Treatment of CNS Tumors

Therapeutic agents have been effectively attached to the surface of gold nanoparticles, which allowed them to be transported into cells. Specifically, gold nanoparticles surface-modified with chemotherapeutics, such as paclitaxel and doxorubicin, have been successfully synthesized and tested [55, 56]. Additionally, doxorubicin-coated gold nanoparticles were able to enter tumor cylindroids [56].

7.3.1.3 Gold Nanoparticles for Treatment of Spinal Cord Injury

Gold nanoparticles may also have a role in the treatment of spinal cord injury. In 2016, Zhang et al. created a gold nanoparticle-based carrier surface-modified with wheat germ agglutinin horseradish peroxidase (WGA-HRP), which concurrently acted as a targeting agent and a visual reporter and either dipropylcyclopentylxanthine or theophylline, drugs that are used as a selective antagonist for adenosine A1 receptor for the treatment of respiratory dysfunctions. WGA-HRP demonstrated the ability to penetrate neurons with adsorptive-mediated endocytosis and to reach neuronal cell bodies by retrograde transport. Although the drug was injected intramuscularly into the diaphragm, in vivo drug release was seen in the cervical spinal cord and medulla nuclei in an experimental rat model [57]. This has potential applications of drug delivery directly targeted into the cervical spinal cord in the context of spinal cord injury.

7.3.1.4   Gold Nanoparticles for Treatment of Glioblastoma Multiforme

Glioblastoma multiforme (GBM) is an aggressive form of malignant glioma. One of the most common adult central nervous system neoplasias, GBM represents nearly 77% of all malignant brain tumors [58]. There are an estimated 25,000 new cases a year in the USA. Current treatments include chemotherapy, radiotherapy and surgical resection [58]. Despite these treatments and advancements in current oncological treatments, the mean survival for GBM patients is only 11 months [59].

Hainfeld et al. were able to create a radiosensitization strategy using gold nanoparticles for brain tumor treatment [54]. Using a mouse model, 1.9-nm-diameter gold nanoparticles injected intravenously were able to preferentially localize into brain gliomas with a 19:1 tumor-to-healthy parenchyma ratio. The relevant accumulation of gold nanoparticles into the tumor tissue enabled a high resolution for tumor imaging by computed tomography and increased sensitivity of the glioma to radiation, prolonging the life the experimental animals when compared to radiation alone [54]. Other studies have explored the use of gold nanoparticles that are surface-modified with polyethylene glycol (PEG) for the treatment of GBM. The BBB can be a hurdle in the clinical treatment of GBM, and in order to circumvent it, PEGylated gold nanoparticles were employed. They were shown to have antibiofouling properties, prolonging systemic circulation half-life and enhancing the efficacy of the treatment. Specifically, these PEGylated gold nanoparticles were tested in conjunction with radiation therapy in cell culture experiments and an animal model. The combination of gold nanoparticles and radiation therapy resulted in localized and specific DNA damage in the GBM tumors [60], demonstrating that gold nanoparticles may not only be useful in carrying chemotherapeutics but can also be used with respect to combination therapy.

7.3.2   Silver Nanoparticles

Silver nanoparticles, shown in Fig. 7.9, (1–100 nm) are similar to gold nanoparticles in the respect that they can also theoretically carry various payloads to specific targets with the assistance of the appropriate surface-modification. Ultra-fine silver

Fig. 7.9 Scanning electron microscopy image of silver nanoparticles 20 nm in size

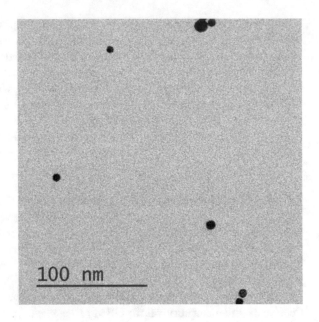

nanoparticles can be inhaled and traverse the blood–lung barrier to move to the circulatory system [61]. However, it has been reported that silver nanoparticles may interact with cerebral microvasculature to produce a proinflammatory cascade and induce the BBB inflammation, astrocyte swelling and neuronal degeneration [62]. The mechanism of this reaction remains to be clarified.

7.3.3 Iron Oxide Nanoparticles

When looking at the overall range of inorganic nanoparticles that have been developed, iron oxide nanoparticles are among the most widely researched. In fact, ferumoxytol, the first iron oxide conjugate, has been approved for limited clinical use by the FDA [63, 64]. Specifically, ultra-small iron oxide nanoparticles are applicable in imaging procedures and utilize their superparamagnetic properties to increase the signal intensity of T1-weighted images and decrease signal intensity of T2*-weighted images in magnetic resonance imaging (MRI) [64]. These properties thereby allow for iron oxide nanoparticles to be used as effective contrast media in neuroimaging platforms [63]. Since iron oxide can be toxic within the biological system, these superparamagnetic nanoparticles have been coated with biocompatible polymers, such as dextran or polyglucose sorbitol carboxymethyl ether, which concurrently allows them to permeate the BBB. One example of such imaging involves intravenous administration of ferumoxytol for the purpose of imaging malignant brain tumors. When compared to conventional gadolinium-based contrast media, ferumoxytol demonstrated higher signal resolution of brain tumor lesions [65, 66].

Other studies have shown effective contrast enhancement with the use of transferrin-conjugated superparamagnetic iron oxide nanoparticles for brain glioma detection [67]. A plethora of other studies include applications in imaging of neuroinflammation and stroke diagnosis. It is obvious that the use of iron oxide nanoparticles presents a promising novel avenue for the diagnosis of a wide variety of CNS disorders.

7.3.3.1 Iron Oxide Nanoparticles in Combination Therapy

Recently, much focus has been directed towards developing nano-agents that can be used for their combined ability to act as imaging and therapeutic agents. One such example was developed by Hu et al., producing a porous iron oxide nanoparticle, loaded with doxorubicin [68]. The porous iron oxide nanoparticles (PIONs) were used as an imaging contrast agent, utilizing the superparamagnetic properties described above. Contrast enhancement was similar to that achieved by the currently used gadolinium agents. Simultaneously, PIONs were used as a photothermal therapy agent. Near-infrared irradiation was utilized after the PIONs were combined with cancer cells, and their ability to convert near-infrared light to heat was assessed. A synergistic effect was achieved when doxorubicin was slowly released through the disrupted walls of the PIONs, which was the result of the overall temperature increase. When comparing the individual effect of the photothermal therapy or doxorubicin to the synergistic effect of the DOX-PIONs, it was evident that the latter formulation was much more effective as a treatment towards these cancer cells [68]. This study demonstrates that nanoparticles can be used to not only carry therapeutic agents as a surface-modification but in combination therapy as hollow and porous carriers, with the carrier shell having its own purpose in the treatment. This combination of properties allows the achievement of an impressive synergistic effect and can increase the effectiveness of the treatment.

7.4 Concluding Remarks

There are many promising pharmaceuticals that can help in the diagnosis and treatment of CNS diseases. However, the BBB presents a physical barrier which limits the use of these therapeutics. Only small, lipophilic drugs can pass through the BBB. This excludes the use of 100% of the large molecule and 98% of small molecule therapeutics [69]. To circumvent this barrier, a variety of physical, chemical, and biological methods of BBB disruption have been employed. However, these methods are far from ideal and cause permanent damage to the BBB with devastating consequences. Nanoparticle-based drug delivery systems present a noninvasive method of drug delivery into the CNS. Nanoparticles are able to carry both lipophilic and lipophobic drugs by entrapping the drugs into cavities, adsorption or conjugation. Nanoparticles are also highly customizable with regard to size, charge and surface functional groups. These advantages should allow for the specific targeting of diseased cells, combination therapy with more than one drug and controlled and sustained release in the CNS. In conclusion, current literature shows that there have been significant advances in the research of nanoparticles for the next generation of CNS pharmaceuticals.

Disclosures The authors have no sources of funding to declare. All authors have read and approved this final version.

References

1. Pardridge WM. Drug targeting to the brain. Pharm Res. 2007;24(9):1733–44.
2. Pardridge WM. Molecular biology of the blood-brain barrier. Mol Biotechnol. 2005;30(1):57–70.
3. Popovic N, Brundin P. Therapeutic potential of controlled drug delivery systems in neurodegenerative diseases. Int J Pharm. 2006;314(2):120–6.
4. Aloe L, Rocco ML, Omar Balzamino B, Micera A. Nerve growth factor: a focus on neuroscience and therapy. Curr Neuropharmacol. 2015;13(3):294–303.
5. Sarkadi B, Homolya L, Szakacs G, Varadi A. Human multidrug resistance ABCB and ABCG transporters: participation in a chemoimmunity defense system. Physiol Rev. 2006;86(4):1179–236.
6. Marchesi VT. The role of pinocytic vesicles in the transport of materials across the walls of small blood vessels. Invest. Ophthalmol. 1965;4(6):1111–21.
7. Zhou Y, Peng Z, Seven ES, Leblanc RM. Crossing the blood-brain barrier with nanoparticles. J Control Release. 2018;270:290–303.
8. Baker D, Gerritsen W, Rundle J, Amor S. Critical appraisal of animal models of multiple sclerosis. Mult Scler. 2011;17(6):647–57.
9. Sloane E, Ledeboer A, Seibert W, Coats B, van Strien M, Maier SF, Johnson KW, Chavez R, Watkins LR, Leinwand L and others. Anti-inflammatory cytokine gene therapy decreases sensory and motor dysfunction in experimental Multiple Sclerosis: MOG-EAE behavioral and anatomical symptom treatment with cytokine gene therapy. Brain Behav. Immun. 2009;23(1):92–100.
10. Male D, Gromnicova R, McQuaid C. Gold nanoparticles for imaging and drug transport to the CNS. Int Rev Neurobiol. 2016;130:155–98.
11. Brodell DW, Jain A, Elfar JC, Mesfin A. National trends in the management of central cord syndrome: an analysis of 16,134 patients. Spine J. 2015;15(3):435–42.
12. Cardoso AM, Guedes JR, Cardoso AL, Morais C, Cunha P, Viegas AT, Costa R, Jurado A, Pedroso de Lima MC. Recent trends in nanotechnology toward CNS diseases: lipid-based nanoparticles and exosomes for targeted therapeutic delivery. Int Rev Neurobiol. 2016;130:1–40.
13. Patel T, Zhou J, Piepmeier JM, Saltzman WM. Polymeric nanoparticles for drug delivery to the central nervous system. Adv Drug Deliv Rev. 2012;64(7):701–5.
14. Lecesne R, Drouillard J, Cisse R, Schiratti M. Contribution of Abdoscan in MRI cholangio-pancreatography and MRI urography. J Radiol. 1998;79(6):573–5.
15. Zong Y, Wu J, Shen K. Nanoparticle albumin-bound paclitaxel as neoadjuvant chemotherapy of breast cancer: a systematic review and meta-analysis. Oncotarget. 2017;8(10):17360–72.
16. Olivier JC, Fenart L, Chauvet R, Pariat C, Cecchelli R, Couet W. Indirect evidence that drug brain targeting using polysorbate 80-coated polybutylcyanoacrylate nanoparticles is related to toxicity. Pharm Res. 1999;16(12):1836–42.
17. Kreuter J, Alyautdin RN, Kharkevich DA, Ivanov AA. Passage of peptides through the blood-brain barrier with colloidal polymer particles (nanoparticles). Brain Res. 1995;674(1):171–4.

18. Schroeder U, Sommerfeld P, Sabel BA. Efficacy of oral dalargin-loaded nanoparticle delivery across the blood-brain barrier. Peptides. 1998;19(4):777 80.
19. Schroeder U, Schroeder H, Sabel BA. Body distribution of 3H-labelled dalargin bound to poly(butyl cyanoacrylate) nanoparticles after i.v. injections to mice. Life Sci. 2000;66(6):495–502.
20. Koffie RM, Farrar CT, Saidi LJ, William CM, Hyman BT, Spires-Jones TL. Nanoparticles enhance brain delivery of blood-brain barrier-impermeable probes for in vivo optical and magnetic resonance imaging. PNAS. 2011;108(46):18837–42.
21. Ramot Y, Haim-Zada M, Domb AJ, Nyska A. Biocompatibility and safety of PLA and its copolymers. Adv Drug Deliv Rev. 2016;107:153–62.
22. Mahapatro A, Singh DK. Biodegradable nanoparticles are excellent vehicle for site directed in-vivo delivery of drugs and vaccines. J. Nanobiotechnology. 2011;9:55.
23. Lu JM, Wang X, Marin-Muller C, Wang H, Lin PH, Yao Q, Chen C. Current advances in research and clinical applications of PLGA-based nanotechnology. Expert Rev. Mol. Diagn. 2009;9(4):325–41.
24. Grabrucker AM, Garner CC, Boeckers TM, Bondioli L, Ruozi B, Forni F, Vandelli MA, Tosi G. Development of novel Zn2 + loaded nanoparticles designed for cell-type targeted drug release in CNS neurons: in vitro evidences. PLoS ONE. 2011;6(3):e17851.
25. Mathew A, Fukuda T, Nagaoka Y, Hasumura T, Morimoto H, Yoshida Y, Maekawa T, Venugopal K, Kumar DS. Curcumin loaded-PLGA nanoparticles conjugated with Tet-1 peptide for potential use in Alzheimer's disease. PLoS ONE. 2012;7(3):e32616.
26. Lei C, Davoodi P, Zhan W, Kah-Hoe Chow P, Wang CH. Development of Nanoparticles for Drug Delivery to Brain Tumor: The Effect of Surface Materials on Penetration into Brain Tissue. Sci: J. Pharm; 2018.
27. Nabeshima T, Nitta A. Memory impairment and neuronal dysfunction induced by beta-amyloid protein in rats. Tohoku J Exp Med. 1994;174(3):241–9.
28. Ringman JM, Frautschy SA, Cole GM, Masterman DL, Cummings JL. A potential role of the curry spice curcumin in Alzheimer's disease. Curr Alzheimer Res. 2005;2(2):131–6.
29. Mulik RS, Monkkonen J, Juvonen RO, Mahadik KR, Paradkar AR. ApoE3 mediated poly(butyl) cyanoacrylate nanoparticles containing curcumin: study of enhanced activity of curcumin against beta amyloid induced cytotoxicity using in vitro cell culture model. Mol Pharm. 2010;7(3):815–25.
30. Doggui S, Sahni JK, Arseneault M, Dao L, Ramassamy C. Neuronal uptake and neuroprotective effect of curcumin-loaded PLGA nanoparticles on the human SK-N-SH cell line. J Alzheimers Dis. 2012;30(2):377–92.
31. Cheng KK, Yeung CF, Ho SW, Chow SF, Chow AH, Baum L. Highly stabilized curcumin nanoparticles tested in an in vitro blood-brain barrier model and in Alzheimer's disease Tg2576 mice. AAPS J. 2013;15(2):324–36.
32. Mathur R, Ince PG, Minett T, Garwood CJ, Shaw PJ, Matthews FE, Brayne C, Simpson JE, Wharton SB. A reduced astrocyte response to beta-amyloid plaques in the ageing brain associates with cognitive impairment. PLoS ONE. 2015;10(2):e0118463.
33. Zhou Y, Sharma S, Peng Z, Leblanc R. Polymers in Carbon Dots: A Review. Polymers. 2017;9(2):67.
34. Li S, Peng Z, Leblanc RM. Method To Determine Protein Concentration in the Protein-Nanoparticle Conjugates Aqueous Solution Using Circular Dichroism Spectroscopy. Anal Chem. 2015;87(13):6455–9.
35. Peng Z, Li S, Han X, Al-Youbi AO, Bashammakh AS, El-Shahawi MS, Leblanc RM. Determination of the composition, encapsulation efficiency and loading capacity in protein drug delivery systems using circular dichroism spectroscopy. Anal Chim Acta. 2016;937:113–8.
36. Xu G, Mahajan S, Roy I, Yong KT. Theranostic quantum dots for crossing blood-brain barrier in vitro and providing therapy of HIV-associated encephalopathy. Front Pharmacol. 2013;4:140.
37. Zheng M, Ruan S, Liu S, Sun T, Qu D, Zhao H, Xie Z, Gao H, Jing X, Sun Z. Self-Targeting Fluorescent Carbon Dots for Diagnosis of Brain Cancer Cells. ACS Nano. 2015;9(11):11455–61.

38. Li S, Amat D, Peng Z, Vanni S, Raskin S, De Angulo G, Othman AM, Graham RM, Leblanc RM. Transferrin conjugated nontoxic carbon dots for doxorubicin delivery to target pediatric brain tumor cells. Nanoscale. 2016;8(37):16662–9.
39. Liu Y, Lu W. Recent advances in brain tumor-targeted nano-drug delivery systems. Expert Opin Drug Deliv. 2012;9(6):671–86.
40. Luciani A, Olivier JC, Clement O, Siauve N, Brillet PY, Bessoud B, Gazeau F, Uchegbu IF, Kahn E, Frija G and others. Glucose-receptor MR imaging of tumors: study in mice with PEGylated paramagnetic niosomes. Radiology 2004;231(1):135–42.
41. DeMarino C, Schwab A, Pleet M, Mathiesen A, Friedman J, El-Hage N, Kashanchi F. Biodegradable Nanoparticles for Delivery of Therapeutics in CNS Infection. J. Neuroimmune Pharmacol. 2017;12(1):31–50.
42. Helm F, Fricker G. Liposomal conjugates for drug delivery to the central nervous system. Pharmaceutics. 2015;7(2):27–42.
43. Pardridge WM. Molecular Trojan horses for blood-brain barrier drug delivery. Curr Opin Pharmacol. 2006;6(5):494–500.
44. McConnell EM, Ventura K, Dwyer Z, Hunt V, Koudrina A, Holahan MR, DeRosa MC. In Vivo Use of a Multi-DNA Aptamer-Based Payload/Targeting System To Study Dopamine Dysregulation in the Central Nervous System. Neurosci: ACS Chem; 2018.
45. Jin SX, Bi DZ, Wang J, Wang YZ, Hu HG, Deng YH. Pharmacokinetics and tissue distribution of zidovudine in rats following intravenous administration of zidovudine myristate loaded liposomes. Pharmazie. 2005;60(11):840–3.
46. Laquintana V, Trapani A, Denora N, Wang F, Gallo JM, Trapani G. New strategies to deliver anticancer drugs to brain tumors. Expert Opin Drug Deliv. 2009;6(10):1017–32.
47. Salvati E, Re F, Sesana S, Cambianica I, Sancini G, Masserini M, Gregori M. Liposomes functionalized to overcome the blood-brain barrier and to target amyloid-beta peptide: the chemical design affects the permeability across an in vitro model. Int. J. Nanomedicine. 2013;8:1749–58.
48. Rueda Dominguez A, Olmos Hidalgo D, Viciana Garrido R, Torres Sanchez E. Liposomal cytarabine (DepoCyte) for the treatment of neoplastic meningitis. Clin Transl Oncol. 2005;7(6):232–8.
49. Khatri P. Evaluation and management of acute ischemic stroke. Continuum (Minneap Minn) 2014;20(2 Cerebrovascular Disease):283–95.
50. Kim H, Britton GL, Peng T, Holland CK, McPherson DD, Huang SL. Nitric oxide-loaded echogenic liposomes for treatment of vasospasm following subarachnoid hemorrhage. Int. J. Nanomedicine. 2014;9:155–65.
51. Hwang H, Jeong HS, Oh PS, Na KS, Kwon J, Kim J, Lim S, Sohn MH, Jeong HJ. Improving Cerebral Blood Flow Through Liposomal Delivery of Angiogenic Peptides: Potential of (1)(8)F-FDG PET Imaging in Ischemic Stroke Treatment. J Nucl Med. 2015;56(7):1106–11.
52. Bernard ED, Beking MA, Rajamanickam K, Tsai EC, Derosa MC. Target binding improves relaxivity in aptamer-gadolinium conjugates. J Biol Inorg Chem. 2012;17(8):1159–75.
53. Du Y, Qin Y, Li Z, Yang X, Zhang J, Westwick H, Tsai E, Cao X. Development of multifunctional nanoparticles towards applications in non-invasive magnetic resonance imaging and axonal tracing. J Biol Inorg Chem. 2017;22(8):1305–16.
54. Hainfeld JF, Smilowitz HM, O'Connor MJ, Dilmanian FA, Slatkin DN. Gold nanoparticle imaging and radiotherapy of brain tumors in mice. Nanomedicine (Lond). 2013;8(10):1601–9.
55. Gibson JD, Khanal BP, Zubarev ER. Paclitaxel-functionalized gold nanoparticles. J Am Chem Soc. 2007;129(37):11653–61.
56. Kim B, Han G, Toley BJ, Kim C-k, Rotello VM, Forbes NS. Tuning payload delivery in tumour cylindroids using gold nanoparticles. Nat. Nanotechnol. 2010;5:465.
57. Zhang Y, Walker JB, Minic Z, Liu F, Goshgarian H, Mao G. Transporter protein and drug-conjugated gold nanoparticles capable of bypassing the blood-brain barrier. Sci. Rep. 2016;6:25794.
58. de Robles P, Fiest KM, Frolkis AD, Pringsheim T, Atta C, St. Germaine-Smith C, Day L, Lam D, Jette N. The worldwide incidence and prevalence of primary brain tumors: a systematic review and meta-analysis. Neuro-Oncology 2015;17(6):776–783.

59. Lara-Velazquez M, Al-Kharboosh R, Jeanneret S, Vazquez-Ramos C, Mahato D, Tavanaiepour D, Rahmathulla G, Quinones-Hinojosa A. Advances in Brain Tumor Surgery for Glioblastoma in Adults. Brain Sci. 2017;7(12):166.
60. Joh DY, Sun L, Stangl M, Al Zaki A, Murty S, Santoiemma PP, Davis JJ, Baumann BC, Alonso-Basanta M, Bhang D and others. Selective targeting of brain tumors with gold nanoparticle-induced radiosensitization. PLoS One 2013;8(4):e62425.
61. Shilo M, Motiei M, Hana P, Popovtzer R. Transport of nanoparticles through the blood–brain barrier for imaging and therapeutic applications. Nanoscale. 2014;6(4):2146–52.
62. Trickler WJ, Lantz SM, Murdock RC, Schrand AM, Robinson BL, Newport GD, Schlager JJ, Oldenburg SJ, Paule MG, Slikker JW and others. Silver Nanoparticle Induced Blood-Brain Barrier Inflammation and Increased Permeability in Primary Rat Brain Microvessel Endothelial Cells. Toxicol Sci. 2010;118(1):160–170.
63. Ajetunmobi A, Prina-Mello A, Volkov Y, Corvin A, Tropea D. Nanotechnologies for the study of the central nervous system. Prog Neurobiol. 2014;123:18–36.
64. Provenzale JM, Silva GA. Uses of nanoparticles for central nervous system imaging and therapy. AJNR Am J Neuroradiol. 2009;30(7):1293–301.
65. Neuwelt EA, Varallyay CG, Manninger S, Solymosi D, Haluska M, Hunt MA, Nesbit G, Stevens A, Jerosch-Herold M, Jacobs PM and others. The potential of ferumoxytol nanoparticle magnetic resonance imaging, perfusion, and angiography in central nervous system malignancy: a pilot study. Neurosurgery 2007;60(4):601–11; discussion 611-2.
66. Neuwelt EA, Varallyay P, Bago AG, Muldoon LL, Nesbit G, Nixon R. Imaging of iron oxide nanoparticles by MR and light microscopy in patients with malignant brain tumours. Neuropathol Appl Neurobiol. 2004;30(5):456–71.
67. Jiang W, Xie H, Ghoorah D, Shang Y, Shi H, Liu F, Yang X, Xu H. Conjugation of functionalized SPIONs with transferrin for targeting and imaging brain glial tumors in rat model. PLoS ONE. 2012;7(5):e37376.
68. Hu Y, Hu H, Yan J, Zhang C, Li Y, Wang M, Tan W, Liu J, Pan Y. Multifunctional Porous Iron Oxide Nanoagents for MRI and Photothermal/Chemo Synergistic Therapy. Bioconjug Chem. 2018;29(4):1283–90.
69. Pardridge WM. The blood-brain barrier: Bottleneck in brain drug development. NeuroRX. 2005;2(1):3–14.

Chapter 8
Nanoparticles for Cornea Therapeutic Applications: Treating Herpes Simplex Viral Infections

Fiona Simpson, François-Xavier Gueriot, Isabelle Brunette and May Griffith

Abstract Herpes Simplex Virus-1 (HSV-1) infections in the eye often originate in the cornea before assuming a latent state in the trigeminal ganglion. During primary infection and upon injury or reactivation, HSV-1 can lead to significant corneal damage. Nanoparticles (NPs) are an emerging strategy for drug delivery to the cornea because they improve the long-term release of anti-HSV-1 drugs, such as nucleoside analogues. Acyclovir, ganciclovir, and valacyclovir have been successfully delivered using both polymer and lipid-based NPs in vitro. Solid silica dioxide NPs have been used to deliver the cathelicidin, LL-37, which prevented HSV-1 infection in corneal epithelial cells. Iron oxide nanoparticles have also been adapted to deliver an anti-HSV-1 DNA vaccine that successfully reduced corneal opacity and HSV-1 markers in a mouse model. Overall, NPs show promise as a delivery method for anti-HSV-1 strategies.

8.1 Introduction

Effective drug delivery to the cornea requires drug release into the tear fluid on the surface of the cornea to treat epithelial conditions and penetrance into the cornea proper to treat deeper tissues [1–3]. Ideally, the drug will accumulate in the aqueous

F. Simpson · I. Brunette · M. Griffith (✉)
Department of Ophthalmology, Faculty of Medicine, Université de Montréal,
C.P. 6128, Succursale Centre-ville, Montréal, QC H3C 3J7, Canada
e-mail: May.Griffith@umontreal.ca

F. Simpson · M. Griffith
Faculty of Medicine, Institute of Biomedical Engineering, Université de Montréal,
C.P. 6128, Succursale Centre-ville, Montréal, QC H3C 3J7, Canada

F. Simpson · I. Brunette · M. Griffith
Centre de Recherche, Hôpital Maisonneuve-Rosemont, 5690 Boulevard Rosemont,
Montréal, QC H1T 2H2, Canada

F.-X. Gueriot
Department of Ophthalmology, Grenoble Alpes University, Boulevard de la Chantourne,
38700 La Tronche, Grenoble, France

© Springer Nature Switzerland AG 2019
E. I. Alarcon and M. Ahumada (eds.), *Nanoengineering Materials for Biomedical Uses*,
https://doi.org/10.1007/978-3-030-31261-9_8

humor located in the anterior chamber between the cornea and the lens, which acts as a reservoir for additional drug release. However, drug delivery to the cornea is still far from optimal. Although eye drops and other topical medications are commonly used, less than five percent of topically applied drugs can reach deeper ocular targets [4]. Thus, many drugs destined for the cornea, e.g., to treat infections, are still administered systemically, to allow a small percentage to reach deeper ocular targets. Nanoparticles (NPs) that can penetrate the tear film or extracellular space between cells have therefore been developed in an attempt to optimize drug delivery to the cornea. NPs developed to date include those delivering therapeutic drugs and growth factors, to those used for diagnostics.

In this chapter, we briefly review the various NP-based therapies that are being developed to treat corneal pathologies. We focus on the treatment of Herpes Simplex Virus (HSV) infections as a model disease, as this is a significant problem worldwide and is the most prevalent cause of infectious corneal blindness in both the developing and developed world.

8.2 Herpes Simplex Virus and Eye Infections

HSV serotype 1 or HSV-1 is the viral strain that causes eye infections. The global prevalence of ocular HSV-1 is estimated at 1.5 million individuals, including 40,000 new cases of monocular visual impairment annually [5]. Ocular herpes is one of the most severe forms of HSV infection. It can lead to a broad panel of ocular damage, including lid ulceration, conjunctivitis, keratitis, anterior uveitis, and rare but severe retinal disease.

The primary herpes infection is asymptomatic in approximately 94% of cases [6]. However, after a phase of replication at the primary infection site (cornea, peri-ocular, or oro-pharyngeal epithelium), the virus enters the corneal nerve endings and gains access to the trigeminal ganglion by retrograde transport. This is followed by a replication phase in this ganglion, after which the virus enters a state of latency, despite the control of the replication by the immune responses [7, 8]. Liedtke et al. found through examination of the trigeminal ganglia from 109 human cadavers at forensic post-mortems that the age-group specific prevalence of HSV neuronal latency increases from 18% in 0–20 years to 100% in persons older than 60 [9]. The virus can then periodically reactivate, particularly after a trigger like immunosuppression (e.g., medically induced, AIDS), surgery, UV or cold exposure. The risk of reactivation is correlated with the latency viral load in the trigeminal ganglion. The latent virus produces progeny that is transported to the periphery of the nerve. The virus then replicates in the cells near the nerve endings. This can lead to asymptomatic viral shedding or symptomatic recurrent infection.

The cornea is the primary site of HSV-1 infection, known as Herpes Simplex Keratitis (HSK) (Fig. 8.1).

HSV-1 corneal infection presents in diverse forms (frequency of appearance reported in the epidemiologic study of Labetoulle et al. [10]): dendritic keratitis

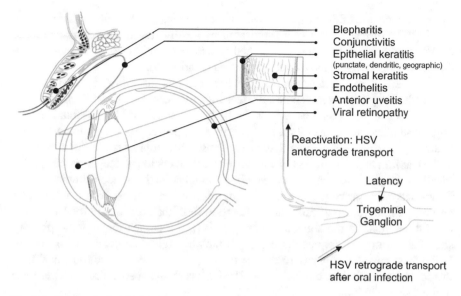

Blepharitis
Conjunctivitis
Epithelial keratitis
(punctate, dendritic, geographic)
Stromal keratitis
Endothelitis
Anterior uveitis
Viral retinopathy

Reactivation: HSV
anterograde transport

Latency

Trigeminal
Ganglion

HSV retrograde transport
after oral infection

Fig. 8.1 Schematic for HSV infection and the different sites of herpes ocular infection

(56.3%), geographic keratitis (9.8%), stromal keratitis (29.5%), punctate keratitis (4%), corneal ulcer (1.8%), limbal lesion (1.1%), and other forms (2.5%) including endothelitis, central corneal oedema, paralimbal interstitial keratitis, and neuroparalytic keratitis. These infections often lead to an irreversible opacification or scarring of the cornea, with scars that result in severe visual impairment. The only treatment to recover functional vision in these eyes is corneal transplantation. However, herpetic recurrences significantly increase the risk of graft failure [5, 11]. In cases of end-stage disease or when the risk for graft failure is too high, an artificial cornea or keratoprosthesis (KPros) becomes an alternative option. This final treatment, however, is overshadowed by severe and frequent side effects, including glaucoma, retroprosthetic membrane formation, infection and dehiscence.

Latency and subsequent reactivation in response to a large number of documented but mechanistically still largely ill-defined triggers make treatment difficult. Nucleoside analogues, including acyclovir (ACV), valacyclovir (VAC) and ganciclovir (GCV), are highly effective at terminating HSV-1 DNA elongation, which inhibits viral proliferation [12]. However, because they target the viral DNA polymerase, they can only act when the virus is in its replicative form. Furthermore, their poor solubility and lack of penetrance across the cornea prevent sufficient accumulation in the anterior segment to efficiently treat ocular HSV-1 infection, which explains the use of oral formulations [13]. But oral administration remains a suboptimal method for treating a local eye infection as a high systemic concentration is needed to yield sufficient drug concentration in the cornea. However, this increases the risk of systemic side effects. Oral administration does not result in rapid accumulation of sufficient levels of nucleoside analogues for the treatment of the local infection.

Systemic exposure of the virus to nucleoside analogues also potentially induces the development of resistance to the antiviral drug. Ideally, new ocular formulations of ACV that address the lack of solubility and corneal penetration would provide a more direct and effective method of treating herpes keratitis [3].

The eye is a particularly challenging field for pharmacology. Several barriers restrict the entry of drug molecules: the tear film, the cornea with its different layers, the conjunctiva, the sclera, as well as the blood-aqueous and blood-retina barriers. Topical administration (eye-drops) is the primary approach for drug delivery in ophthalmology, being used in more than 90% of cases. The topical treatment is released in the tear film and penetrates through the corneal layers to the deeper tissues [1–3]. Ideally, a drug penetrating the cornea will accumulate in the aqueous humor, which in turn acts as a reservoir for additional drug release and treatment of the intraocular tissues. Despite its significant barrier function (less than 5% of the total administered doses reaches the aqueous humor [4, 14]), the cornea remains the primary site of absorption for intraocular drugs. To facilitate release kinetics, nanoparticles (NPs) must have the capacity to release the drug into aqueous solutions, penetrate the corneal epithelial, stromal and endothelial layers and sustain drug release once in the aqueous humor.

We summarize below the development of encapsulated formulations of nucleoside analogues for ocular administration from the first liposomal formulations developed in the late 1980s to current developments in nanoparticle formulations designed to prevent HSV-1 infections.

8.3 History of Encapsulation of Acyclovir in the Eye

8.3.1 Liposomes

Norley et al. developed the first encapsulated form of ACV in 1986 [15, 16]. Using a liposomal formulation targeted with HSV glycoprotein D (gD), they were able to demonstrate binding to human corneal epithelial cells (HCECs) infected with HSV-1 and inhibition of replication at a concentration of 10 µg/mL of ACV.

In 1999, Fresta et al. examined the impact of phospholipid type and liposomal charge on the efficacy of ACV delivery [17]. Their results demonstrated that negatively charged dipalmitoylphosphatidylcholine-cholesterol-dimethyldioctadecyl glycerol bromide liposomes synthesized by reverse-phase evaporation were the most effective, with the highest ACV content. In rabbits, this formulation increased the ACV concentration in the aqueous humor from 2.11 ± 1.80 µg/mL with soluble ACV to 88.95 ± 12.31 µg/mL. The effect of liposomal charge was re-examined by Law et al. in 2000 [18]. Liposomes composed of phosphatidylcholine, cholesterol, stearylamine or dicetylphosphate containing ACV were synthesized with either a positive or negative charge. In rabbits, the concentration of ACV was highest in the corneal tissue and aqueous humor after the administration of positively charged liposomes. These results indicated that lipid carriers are effective at resolving the solubility issues associated with nucleoside analogues penetrating the cornea and paved the way for future lipid nanoparticle formulations.

8.3.2 Chitosan

In 1997, Genta et al. developed the first chitosan microparticle formulation of ACV [19]. The particles were 25 μm in diameter. In an in vivo test in rabbits, the microsphere (MS) formulation was able to increase the concentration of ACV in the aqueous humor in comparison to ACV alone, and it doubled the length of time that ACV was present in the aqueous humor.

8.4 Ocular Nanoparticles for Nucleoside Analogues

8.4.1 Polymer-Based Nanoparticles

8.4.1.1 Poly-(D,L-Lactic Acid)

Poly-(D,L-Lactic Acid) (PLA) NPs were developed by Giannavola et al. to test the effect of poly(ethylene glycol) (PEG) coating on ACV release [20]. NPs containing PLA of three molecular weights (LMW—16,000; MMW—109,000; HMW—209,00) were synthesized. Drug release assays demonstrated that lower molecular weight PLA resulted in greater ACV release. The LMW-PLA nanoparticles coated with PEG resulted in improved drug release over the PLA-only NPs. In normal rabbit eyes, the PEG-PLA encapsulated ACV NPs resulted in the highest concentration of ACV (424 $\pm$ 24 μg/mL/min) released into the aqueous humor. This effect was decreased to the level of PLA-only NPs with the pre-administration of the topical irritant N-acetylcysteine (221 $\pm$ 40 μg/mL/min), which suggests that the formulation may be less effective in an inflamed eye. All NP formulations had greater aqueous humor concentration than soluble ACV.

8.4.1.2 Poly-(Lactic-Co-glycolic Acid)

Jwala et al, developed poly-(lactic-co-glycolic acid) (PLGA) NPs based on ACV prodrugs [21]. PLGA is a US FDA-approved biodegradable polymer that has been used extensively in micro-capsule formulations [22]. The prodrugs were synthesized with the addition of L- and D- valine. LL-ACV and LD-ACV showed the highest degradation rate constants of log[[pro-drug concentration]] versus time in cultured cells and rabbit ocular tissues, and were used in subsequent experiments. PLGA NPs with varying LA:GA ratios (100:0, 75:25, 65:35, 50:50) were tested for encapsulation of the ACV prodrugs. LL-ACV PLGA 75:25 NPs and LD-ACV PLGA 65:35 NPs showed the greatest entrapment efficiency (EE) and drug content. In vitro, the LL-ACV NPs showed a greater controlled release time than the LD-ACV NPs. The release of both prodrugs was extended after incorporation within PLGA-PEG-PLGA thermosensitive gels.

Yang et al. applied the methodology of Jwala et al. to PLGA NPs with GCV prodrugs [23, 24]. They first developed NPs containing LL-GCV, LD-GCV, and DL-GCV with sizes ranging from 116 to 143 nm in diameter and EE of 38.7–45.3% [23]. As these formulations showed successful biphasic drug release and did not demonstrate cytotoxicity, they synthesized a fluorescein isothiocyanate (FITC) PLGA conjugate to visualize cellular uptake of the NPs into human corneal epithelial cells (HCECs) [24]. The FITC-PLGA-NPs ranged from 115 to 145 nm with zeta-potentials around -13 mV. Fluorescent microscopy demonstrated the successful uptake of the NPs by the HCECs indicating that they could successfully penetrate the corneal epithelium.

8.4.1.3 Chitosan

Calderon et al. compared chitosan microspheres (CMS) and nanoparticle (CNPs) of ACV encapsulated in chitosan [25]. Chitosan NPs crosslinked with tripolyphosphate (TPP) were prepared using ionotropic gelation. The CNPs were 240.0 ± 62.4 nm in diameter, while the CMS was 6.2 ± 0.5 μm. The CMS had a higher EE (75.46%) than the CNPs (15.73%), but both had positive zeta-potentials. Both spheres and particles released approximately 75% of the encapsulated drug over 25 h in solution for a total of 430 and 80 μg for CMPs and CNPs, respectively. Due to the comparatively lower ACV release from the CNP, only the CMS were assayed for slug mucosal irritation, which is used to simulate ocular irritation, where they showed a moderate level of irritation.

8.4.1.4 Polycaprolactone

Ramyadevi et al. developed polycaprolactone (PCL) NPs encased in a thermosensitive Pluronic F-127 and Carbopol gel [26]. The NPs were synthesized using solvent evaporation using varying amounts of PCL and concentrations of surfactant. The resulting NPs were 172–329 nm in diameter with an EE of 28–58%. Drug release from the gel varied from 7 to 32% of the encapsulated drug. The formulation PCL3 composed of 100 mg PCL and 0.5% Pluronic F-127 with a size of 201.4 nm and an EE of 64% was used to develop the thermosensitive gel formulations as it showed the greatest sustained release and no cytotoxicity. The gel formulation with the best performance was synthesized using 15% w/v Pluronic F-127 and 0.3% Carbopol 940. The gel showed a linear drug release of 4% for the first hour and 14% at 8 h, following Korsmeyer-Peppas kinetics.

8.4.1.5 Solid Prodrug Nanoparticles

Stella et al. created a novel prodrug by conjugating ACV to 1,10,2-trisnorsqualenoic acid resulting in 40-trisnorsqualenoylacyclovir (SQ-ACV) [27]. The SQ-ACV NPs were formed by spontaneous precipitation resulting in NPs ranging from 113 to 254 nm in diameter with a polydispersity index <0.1 and a negative zeta-potential. After topical administration to rabbits, the SQ-ACVNPs had a greater concentration of total ACV in the tear fluid (11 $\perp$ 2.06 mg/ml min $^{-1}$ vs. 2.97 $\pm$ 0.34 mg/ml min^{-1}) than soluble ACV. The ACV concentration was higher for the SQACV NPs at 1 h (0.294 $\pm$ 0.119 μg/mL) than soluble ACV (0.098 $\pm$ 0.051 μg/mL).

8.4.2 Lipid Nanoparticles

8.4.2.1 Nanostructured Lipid Carriers

Seyfoddin et al. compared nanostructured lipid carriers (NLC) to solid lipid nanoparticles (SLN) in 2013 [28]. SLNs were synthesized as microemulsions from varying amounts of Compritol alone, in combination with stearic acid or cithrol GMS using the hot oil-in-water method in the presence of Tween-20. The Compitrol SLNs were optimized using different types and concentrations of surfactants, resulting in a 400 mg Compritol SLN containing 40 mg ACV synthesized using 2% Tween-40 (SLN3). SLN3 was 465.86 $\pm$ 7.15 nm in size with a negative zeta-potential and a polydispersity index (PDI) of 0.530 $\pm$ 0.05. The NLCs were synthesized using 400 mg of Comptriol containing Capryol-90 and/or Lauroglycol-90 in the presence of Tween-40. The resulting NPs were 319–656 nm in size with negative zeta-potential and PDI from 0.265 to 0.752. In vitro drug release using a diffusion chamber was similar for SLN3, NLC4, and NLC5. The penetration of the NPs was measured using ex vivo bovine corneas in a Franz-type diffusion chamber. NLC5 showed the greatest ACV concentration through the cornea, followed by NLC4. Interestingly, SLN3 showed lower corneal penetration than free ACV.

8.4.2.2 Solid Lipid Nanoparticles

Valacyclovir (VAC) is a prodrug that is converted into ACV within the patient's body. SLNs containing VAC were developed by Kumar et al. [29]. The SLNs were generated using emulsification/evaporation of stearic acid or tristearin, using poloxamer and sodium taurocholate as surfactants, in different rations of lipid:surfactant. The SLNs produced had a particle size from 202.5 to 431.7 nm. The PDI was 0.252 $\pm$ 0.06 and 0.598 $\pm$ 0.03 with all particles other than SLN-2 demonstrating a PDI > 0.5. In vitro drug release assays showed a total release of 60% of the encapsulated drug. The kinetics showed an initial burst release followed by sustained release. Ex vivo corneal penetration assays demonstrated that SLN-4 and SLN-6 had the highest corneal flux, while SLN-3, -4 and -6 had the highest corneal retention (Fig. 8.2).

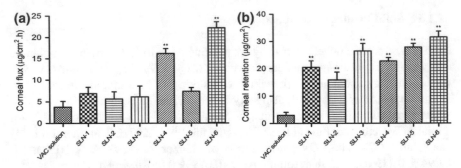

Fig. 8.2 Ex vivo corneal permeation of VAC formulations (**$P \leq 0.01$; statistically very significant as analyzed by Dunnett multiple comparison test). Reproduced with permission from Kumar et al. [29]

Irritation assays were conducted for SLN-6 using a Hen's Egg Test Chorio Allantoic Membrane (HET-CAM) assay with 0.1 N sodium hydroxide as a positive control. SLN-6 did not result in lysis, hemorrhage or coagulation, with the same HET-CAM score as the saline control. Histopathology of SLN-6 in an ex vivo goat corneal model showed morphology similar to untreated controls.

8.5 Experimental Ocular Nanoparticles for Delivery of Nucleoside Alternatives

More recently, cationic peptides have been developed as alternatives to nucleosides for inhibiting HSV infections. Amongst the various peptides examined are the innate cationic peptides such as cathelicidins and defensins produced by the cornea [30]. Humans produce only one cathelicidin, of which a 37-amino acid sequence was found to have anti-viral activity against HSV. This peptide, known as LL-37, is highly cationic and has been shown to have anti-HSV-1 activity. LL-37, its derivatives and other innate peptides are being proposed as new treatments and therefore also require optimal delivery to the cornea. Other cationic peptides with anti-HSV activity include the entry blocker (EB) peptide, which consists of the fibroblast growth factor 4 (FGF4) signal sequence with an additional N-terminal RRKK sequence [31].

Lee et al. used silica dioxide NPs to encapsulate LL-37 and EB for delivery to corneal epithelial cells, using an immortalized line of human epithelial cell as a model line [32]. The LL-37 containing NPs were further encapsulated within a collagen- 2-methacryloyloxyethyl phosphorylcholine (collagen-MPC) hydrogel. LL-37 released from the NPs in implants blocked HSV-1 infection of HCECs more effectively than EB. Controls consisted of free LL-37 within the hydrogels, where the peptides diffused out rapidly and had little effect. The released LL-37 blocked HSV-1 by interfering with viral binding (Fig. 8.3a, c). However, in pre-infected HCECs, LL-37 delayed but did not prevent viral spreading nor clear viruses from the infected cells (Fig. 8.3b, d).

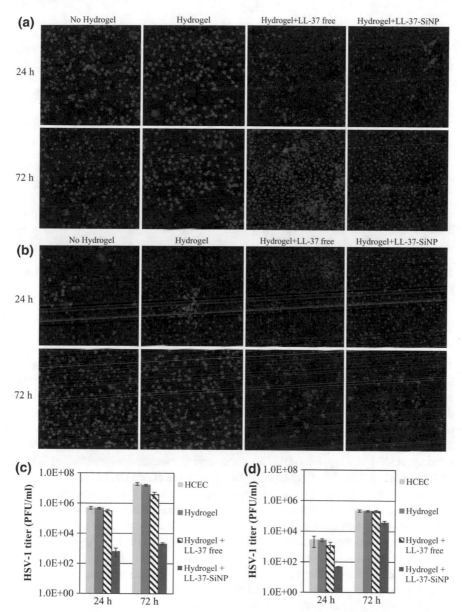

Fig. 8.3 **a** The prophylactic effects of collagen-MPC hydrogels that incorporated LL-37 SiNPs, free LL-37, or nothing in HSV-1 infection (MOI ¼ 0.1). Red immunofluorescence indicated localization of HSV-1 infected cells; cell nuclei are stained with DAPI (blue). **b** Effects of LL-37 release on virus spreading, after inoculation with HSV-1 (MOI ¼ 0.05). **c** Quantification of HSV-1 titers from (**a**). **d** Quantification of HSV-1 titers corresponding (**b**). Reproduced with permission from Lee C-J et al. [32]

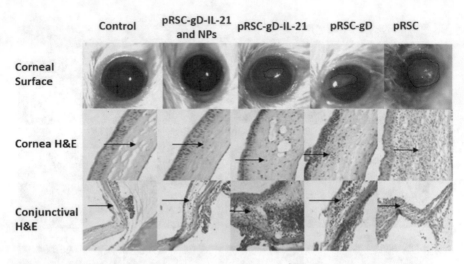

Fig. 8.4 Corneal and conjunctival response to HSV- 1 challenge after vaccination. Adapted from Figures 5 and 6, Hu et al. [34]

8.6   Ocular Nanoparticles for Herpes Vaccine Delivery

There is currently no available vaccine against HSV infection [33]. A number of studies published from 1980s to recent times showed that vaccine development attempts were largely unsuccessful. In 2011, Hu et al. used iron oxide (Fe_3O_4) NPs to improve the delivery of a DNA vaccine against HSV-1 [34]. The DNA vaccine consists of a pRSC vector with cDNA for both HSV glycoprotein D (gD) and interleukin-21 (IL-21). This vector was co-administered with Fe_3O_4 NPs coated with glutamic acid in a mouse HSV-1 challenge model. The mice were immunized with pRSC-gD-IL-21 with and without NPs, pRSC-gD, and pRSC three times at two weeks intervals. Quantification of the vaccine efficacy was conducted by measuring secreted IgA, serum neutralizing antibody, IL-4, and IFN-γ. For all measures, DNA + NPs improved the expression of markers over the DNA vaccine alone. Three weeks after the final immunization, the mice were infected with HSV-1 and followed for 15 days. Figure 8.4 shows the clinical appearance of the cornea at follow up and histopathology of the cornea and conjunctiva. The vaccine-containing NP group had minimal corneal opacity and normal corneal and conjunctival structure, but the vaccine only or vector only groups showed corneal opacity and structural damage.

8.7   Outlook and Future Perspective

A wide variety of nanoparticles have been used to improve the delivery of nucleoside analogues to the cornea; however, these investigations are limited to early in vivo drug

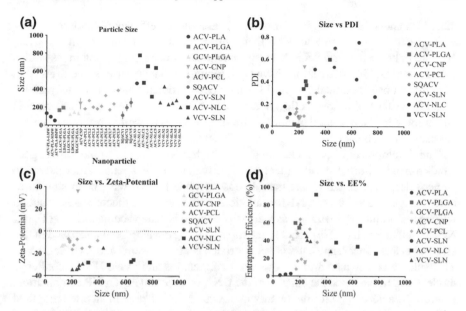

Fig. 8.5 Summary of nucleoside analogue nanoparticle properties. **a** Nanoparticle size (nm). **b** Size (nm) versus polydispersity index (PDI). **c** Nanoparticle size versus zeta-potential (mV). **d** Nanoparticle size versus entrapment efficiency (%). Original data obtained from cited literature in this chapter

delivery and toxicology studies. There is considerable variability in the relationship between NP size and polydispersity index (PDI), zeta-potential, and entrapment efficiency (EE). While size versus PDI, and size versus EE can be loosely approximated with a linear relationship, where increased size results in increased polydispersity and EE, there is no obvious relationship between size and zeta-potential (Fig. 8.5). Chitosan is distinct from the other polymeric NPs in both zeta-potential and morphology. Where most polymeric NPs have a negative zeta-potential (Fig. 8.5) and spherical shape, the CNPs have a positive zeta-potential and red blood cell-like appearance. The SLNs have a smooth surface that is more reminiscent of a polymeric NPs, while the NLCs have highly textured surfaces. Like most polymeric NPs, the NLCs have a negative zeta-potential but their larger size and unique surfaces make their increased penetration across the cornea (in comparison to SLNs) more likely to be dependent on their hydrophobic properties than their particle size. This suggests that smaller NPs are not always superior drug carriers unless they have equally effective particle design.

It is unfortunate that researchers have used highly variable methodologies and units to compare NP diffusion across the cornea, making it unclear which NPs are the most effective drug delivery vehicles. The drug release studies have used a more consistent methodology, with several formulas showing a high total release, but this measure is not indicative of overall NP performance in vivo. It is also problematic that researchers developing NP formulations assume the efficacy of nucleoside analogues without conducting tests on their antiviral efficacy in vitro or in vivo, i.e., NPs are only the delivery vehicle. Resistance to ACV is a rising problem, as strains with thymidine kinase mutations have been identified that are insensitive to ACV in immunocompetent clinical patients [35] and hence, problems have arisen for patients who are prescribed prophylactic systemic doses of ACV or VAC. It is critical that NP formulations are studied in clinical strains of HSV-1, as well as reference strains, to ensure their efficacy.

The development of NPs that can deliver high doses directly to the cornea goes hand-in-hand with the development of effective anti-HSV agents. Alternatives to nucleosides, such as the innate host defense peptide LL-37, have been studied as potential anti-HSV-1 agents. NP formulations with alternative mechanisms of action will be important for treatment, as well as prophylaxis. The vaccine utilizing Fe_3O_4 NPs as delivery vehicles is probably the most exciting development described in this article, because they represent a novel approach for the prevention of ocular HSV-1. This study was also a comprehensive one that included an in vivo HSV-1 challenge model that demonstrated the efficacy of the DNA vaccine and Fe_3O_4 NP combination. Overall, NPs developed for the treatment of ocular HSV-1 remain an emerging field with a great deal of potential for drug development. The standardization of NP assessment has improved the ability to compare the properties of the NP formulations, but a standard in vivo model for corneal diffusion and efficacy would greatly improve the ability of researchers to determine if new formulations are improvements over existing NPs.

Acknowledgements The authors have no conflict of interest. FS is supported by a FRQNT PhD studentship. FXG is supported by a (Berthe Fouassier) France Foundation studentship. MG holds the Caroline Durand Foundation Research Chair for Cellular Therapy of Diseases of the Eye, Université de Montréal.

Disclosure All authors have read and approved the final version.

References

1. Agarwal P, Rupenthal ID. In vitro and ex vivo corneal penetration and absorption models. Drug Deliv Transl Res. 2016;6(6):634–47.
2. Reimondez-Troitino S, Csaba N, Alonso MJ, de la Fuente M. Nanotherapies for the treatment of ocular diseases. Eur J Pharm Biopharm. 2015;95(Pt B):279–93.
3. Sharma A, Taniguchi J. Review: Emerging strategies for antimicrobial drug delivery to the ocular surface: implications for infectious keratitis. Ocul Surf. 2017;15(4):670–9.
4. Gaudana R, Jwala J, Boddu SH, Mitra AK. Recent perspectives in ocular drug delivery. Pharm Res. 2009;26(5):1197–216.
5. Farooq AV, Shukla D. Herpes simplex epithelial and stromal keratitis: an epidemiologic update. Surv Ophthalmol. 2012;57(5):448–62.
6. Liesegang TJ, Melton LJ III, Daly PJ, Ilstrup DM. Epidemiology of ocular herpes simplex. Incidence in Rochester, Minn, 1950 through 1982. Arch Ophthalmol. 1950;107(8):1155–9.

7. Kennedy DP, Clement C, Arceneaux RL, Bhattacharjee PS, Huq TS, Hill JM. Ocular herpes simplex virus type 1: is the cornea a reservoir for viral latency or a fast pit stop? Cornea. 2011;30(3):251–9.
8. Al-Dujaili LJ, Clerkin PP, Clement C, McFerrin HE, Bhattacharjee PS, Varnell ED, Kaufman HE, Hill JM. Ocular herpes simplex virus: how are latency, reactivation, recurrent disease and therapy interrelated? Future Microbiol. 2011;6(8):877–907.
9. Liedtke W, Opalka B, Zimmermann CW, Lignitz E. Age distribution of latent herpes simplex virus 1 and varicella-zoster virus genome in human nervous tissue. J Neurol Sci. 1993;116(1):6–11.
10. Labetoulle M, Auquier P, Conrad H, Crochard A, Daniloski M, Bouee S, El Hasnaoui A, Colin J. Incidence of herpes simplex virus keratitis in France. Ophthalmology. 2005;112(5):888–95.
11. Fry M, Aravena C, Yu F, Kattan J, Aldave AJ. Long-term outcomes of the Boston type I keratoprosthesis in eyes with previous herpes simplex virus keratitis. Br J Ophthalmol. 2018;102(1):48–53.
12. De Clercq E, Holy A. Acyclic nucleoside phosphonates: a key class of antiviral drugs. Nat Rev Drug Discov. 2005;4(11):928–40.
13. Sanchez-Lopez E, Espina M, Doktorovova S, Souto EB, Garcia ML. Lipid nanoparticles (SLN, NLC): overcoming the anatomical and physiological barriers of the eye—Part II—Ocular drug-loaded lipid nanoparticles. Eur J Pharm Biopharm. 2017;110:58–69.
14. Hughes PM, Olejnik O, Chang-Lin JE, Wilson CG. Topical and systemic drug delivery to the posterior segments. Adv Drug Deliv Rev. 2005;57(14):2010–32.
15. Norley SG, Huang L, Rouse BT. Targeting of drug loaded immunoliposomes to herpes simplex virus infected corneal cells: an effective means of inhibiting virus replication in vitro. J Immunol. 1986;136(2):681–5.
16. Norley SG, Sendele D, Huang L, Rouse BT. Inhibition of herpes simplex virus replication in the mouse cornea by drug containing immunoliposomes. Invest Ophthalmol Vis Sci. 1987;28(3):591–5.
17. Fresta M, Panico AM, Bucolo C, Giannavola C, Puglisi G. Characterization and in-vivo ocular absorption of liposome-encapsulated acyclovir. J Pharm Pharmacol. 1999;51(5):565–76.
18. Law SL, Huang KJ, Chiang CH. Acyclovir-containing liposomes for potential ocular delivery. Corneal penetration and absorption. J Control Release. 2000;63(1–2):135–40.
19. Genta I, Conti B, Perugini P, Pavanetto F, Spadaro A, Puglisi G. Bioadhesive microspheres for ophthalmic administration of acyclovir. J Pharm Pharmacol. 1997;49(8):737–42.
20. Giannavola C, Bucolo C, Maltese A, Paolino D, Vandelli MA, Puglisi G, Lee VH, Fresta M. Influence of preparation conditions on acyclovir-loaded poly-d, l-lactic acid nanospheres and effect of PEG coating on ocular drug bioavailability. Pharm Res. 2003;20(4):584–90.
21. Jwala J, Boddu SH, Shah S, Sirimulla S, Pal D, Mitra AK. Ocular sustained release nanoparticles containing stereoisomeric dipeptide prodrugs of acyclovir. J Ocul Pharmacol Ther. 2011;27(2):163–72.
22. Makadia HK, Siegel SJ. Poly Lactic-co-glycolic acid (PLGA) as biodegradable controlled drug delivery carrier. Polymers. 2011;3(3):1377–97.
23. Yang X, Shah SJ, Wang Z, Agrahari V, Pal D, Mitra AK. Nanoparticle-based topical ophthalmic formulation for sustained release of stereoisomeric dipeptide prodrugs of ganciclovir. Drug Deliv. 2016;23(7):2399–409.
24. Yang X, Sheng Y, Ray A, Shah SJ, Trinh HM, Pal D, Mitra AK. Uptake and bioconversion of stereoisomeric dipeptide prodrugs of ganciclovir by nanoparticulate carriers in corneal epithelial cells. Drug Deliv. 2016;23(7):2532–40.
25. Calderon L, Harris R, Cordoba-Diaz M, Elorza M, Elorza B, Lenoir J, Adriaens E, Remon JP, Heras A, Cordoba-Diaz D. Nano and microparticulate chitosan-based systems for antiviral topical delivery. Eur J Pharm Sci. 2013;48(1–2):216–22.
26. Ramyadevi D, Sandhya P. Dual sustained release delivery system for multiple route therapy of an antiviral drug. Drug Deliv. 2014;21(4):276–92.
27. Stella B, Arpicco S, Rocco F, Burgalassi S, Nicosia N, Tampucci S, Chetoni P, Cattel L. Non-polymeric nanoassemblies for ocular administration of acyclovir: pharmacokinetic evaluation in rabbits. Eur J Pharm Biopharm. 2012;80(1):39–45.

28. Seyfoddin A, Al-Kassas R. Development of solid lipid nanoparticles and nanostructured lipid carriers for improving ocular delivery of acyclovir. Drug Dev Ind Pharm. 2013;39(4):508–19.
29. Kumar R, Sinha VR. Lipid nanocarrier: an efficient approach towards ocular delivery of hydrophilic drug (valacyclovir). AAPS Pharm Sci Tech. 2017;18(3):884–94.
30. Gordon YJ, Huang LC, Romanowski EG, Yates KA, Proske RJ, McDermott AM. Human cathelicidin (LL-37), a multifunctional peptide, is expressed by ocular surface epithelia and has potent antibacterial and antiviral activity. Curr Eye Res. 2005;30(5):385–94.
31. Bultmann H, Busse JS, Brandt CR. Modified FGF4 signal peptide inhibits entry of herpes simplex virus type 1. J Virol. 2001;75(6):2634–45.
32. Lee CJ, Buznyk O, Kuffova L, Rajendran V, Forrester JV, Phopase J, Islam MM, Skog M, Ahlqvist J, Griffith M. Cathelicidin LL-37 and HSV-1 corneal infection: peptide versus gene therapy. Transl Vis Sci Technol. 2014;3(3):4.
33. Johnston C, Gottlieb SL, Wald A. Status of vaccine research and development of vaccines for herpes simplex virus. Vaccine. 2016;34(26):2948–52.
34. Hu K, Dou J, Yu F, He X, Yuan X, Wang Y, Liu C, Gu N. An ocular mucosal administration of nanoparticles containing DNA vaccine pRSC-gD-IL-21 confers protection against mucosal challenge with herpes simplex virus type 1 in mice. Vaccine. 2011;29(7):1455–62.
35. Piret J, Boivin G. Antiviral resistance in herpes simplex virus and varicella-zoster virus infections: diagnosis and management. Curr Opin Infect Dis. 2016;29(6):654–62.

Chapter 9
Therapeutic Use of Bioengineered Materials for Myocardial Infarction

Veronika Sedlakova, Marc Ruel and Erik J. Suuronen

Abstract Cardiovascular disease is a leading cause of worldwide mortality. Despite the success of current therapies for acute myocardial infarction (MI), many patients still suffer irreversible damage, and the prevalence of heart failure is growing. After MI, the extracellular matrix (ECM) of the damaged myocardium is modified to produce scar tissue. This remodeling reduces the efficacy of therapies and also hinders endogenous repair mechanisms. Therefore, a strategy to prevent adverse remodeling and provide a suitable ECM environment that supports cells, tissue repair and functional restoration may lead to a superior therapeutic outcome in MI patients. Bioengineered materials are an attractive approach for achieving this. Herein, we review current research on materials that can act as a biomimetic matrix for supporting cellular repair in the post-MI heart. We also examine how nanomaterials are being used to treat the damaged heart. Finally, we provide an overview of the breakthroughs and limitations of biomaterial therapies for cardiac repair.

9.1 Introduction

Ischemic heart disease (IHD), accountable for over 9 million deaths in 2016, is a leading cause of mortality worldwide [1]. It is characterized by a lack of oxygen to meet the supply requirement of the myocardium [2]. Based on clinical symptoms, the manifestation ranges from acute forms (angina pectoris, myocardial infarction, sudden cardiac death) to chronic ones (ischemic cardiomyopathy) [2–4].

Acute myocardial infarction (MI) affects approximately 7 million people annually [5, 6] and is among the deadliest forms of IHD [2]. Current diagnostic options based on early recognition of typical symptoms, elevated biomarkers of cardiomyocyte necrosis and hallmark electrocardiographic changes [3, 6] help to commence the therapeutic process as early as possible. Despite this, loss of the myocardium is simply inevitable. In the best-case scenario of early reperfusion, the damage may be limited

V. Sedlakova · M. Ruel · E. J. Suuronen (✉)
BEaTS Research Program, Division of Cardiac Surgery, University of Ottawa Heart Institute, 40 Ruskin Street, Ottawa, ON K1Y 4W7, Canada
e-mail: esuuronen@ottawaheart.ca

just to the subendocardial layer [3]. However, with prolonged ischemia and increased size of the affected area, outcomes are more severe and correlate with both early and late patient survival. The patient is also at risk of additional complications such as arrhythmias (atrial and ventricular fibrillation, atrioventricular blocks), ruptures of papillary muscle, septum or free wall, pericarditis, stroke, cardiogenic shock [6] or heart failure [3]. The risk of progression to heart failure is proportional to the extent of adverse ventricular remodeling after MI [2, 3].

In the failing heart, the heart´s ability to pump and/or fill with blood is severely compromised [7]. Initial heart failure therapies rely largely on pharmacological relief of the symptoms [8], followed by management of comorbidities, physiotherapy or medical device implantation. Nevertheless, the prognosis is very poor with an approximately 30% one-year mortality [7, 8]. For the end-stage failing heart, few treatment options exist [9] and consist of the implantation of mechanical ventricular assist devices and possibly heart transplantation, although donor organs are in short supply. Consequently, there is an urgent need to develop strategies that can effectively limit cardiac remodeling and promote cardiac repair and thus prevent the progression to heart failure.

9.2   Cardiac Extracellular Matrix

The human heart is composed of endocardium, myocardium and epicardium and is enclosed within a pericardial sac [10]. The myocardium is comprised of cardiomyocytes surrounded by cardiac interstitium. The interstitium contains various other cell types (fibroblasts, endothelial cells, macrophages, lymphocytes and some adipocytes) and cardiac extracellular matrix (ECM) providing essential support to the contractile tissue. The cardiac ECM contains key structural proteins (mainly collagens) surrounded by glycoproteins and proteoglycans [11].

Collagen type I is the most abundant type of collagen in the heart (~80%), followed by type III (~10%) [12]. Collagen type I provides the bulk of the structural support to the myocardium, while collagen type III is primarily responsible for conferring its elasticity [13, 14]. The myocardium also contains lesser amounts of collagen type IV, V and VI [14]. Cells interact with specific sequences present on collagen molecules via integrin receptors [15], which in turn triggers various intracellular signaling pathways that regulate cell functions including adhesion, migration, proliferation, differentiation or apoptosis [14, 16].

As for other ECM constituents, elastin is present to provide elasticity to the heart muscle [12]. Laminin and fibronectin act as linkers between cells and other ECM components, as they also contain integrin recognition motifs [12, 14–17]. Thrombospondin and tenascin also mediate cell–ECM interactions [17], and together with osteonectin, osteopontin and periostin play important roles in post-MI healing [11, 18].

Proteoglycans found in the cardiac ECM include decorin, perlecan and biglycan, along with the glycosaminoglycan hyaluronic acid [11]. These molecules bind water, participate in post-MI healing both structurally and functionally (by activation/inhibition of cells/pathways) and also have roles in development.

ECM turnover is precisely regulated by various proteases and their inhibitors [19]. For example, matrix metalloproteinases (MMPs) are produced both under normal and pathological conditions and regulate not only ECM degradation, but also cell growth, survival or angiogenesis [19, 20]. Their role post-MI will be discussed later in this chapter.

Taken together, the cardiac ECM is a precisely organized, yet highly dynamic structure, which plays important roles in tissue homeostasis in both health and disease [11, 18].

9.3 Myocardial Infarction is Healed in Three Phases

Myocardium damaged by an ischemic event needs to be repaired in order to effectively maintain the heart's function as a pump and thus maintain oxygen supply to the body. However, the low inherent regenerative potential of the heart significantly limits post-MI repair [2]. The infarcted area undergoes a series of changes ultimately resulting in scar formation. The healing process can be divided into three overlapping phases: the inflammatory, proliferative and maturation phases (Fig. 9.1) [2, 21].

The phases overlap in their duration [2, 22]. It can be approximated that in humans, the inflammatory phase lasts from the time of the ischemic event up to four days after MI, the proliferative phase from the event up to three weeks and the maturation phase from the second to the sixth week post-MI [22].

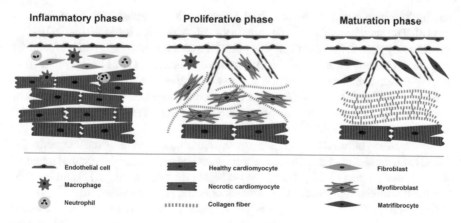

Inflammatory phase	Proliferative phase	Maturation phase

●—●	Endothelial cell	▬▬	Healthy cardiomyocyte	◆	Fibroblast
✱	Macrophage	▬▬	Necrotic cardiomyocyte	◆◆	Myofibroblast
◉	Neutrophil	∷∷∷∷	Collagen fiber	◆	Matrifibrocyte

Fig. 9.1 Phases of healing post-myocardial infarction

9.3.1   The Inflammatory Phase

Following an ischemic event, necrotic cardiomyocytes and damaged extracellular matrix (ECM) produce danger-associated molecular patterns (DAMPs) which trigger an inflammatory response [21]. Upregulation of cytokines, chemokines and cell adhesion molecules together with increased vascular permeability due to hypoxia result in neutrophil, lymphocyte and monocyte infiltration as well as platelet activation [2, 21]. Resident mast cells, macrophages and fibroblasts are also stimulated [21]. This cascade of events creates an inflammatory environment which is essential for clearing the affected area of its cellular and matrix debris.

Evidence suggests that cardiac ECM is not only a passive player in this process [18]. Latent and later newly synthesized MMPs are quickly activated following MI. They digest the cardiac ECM into low molecular weight matrix fragments with pro-inflammatory and other signaling properties. Both MMPs and these matrix fragments then actively regulate the inflammatory reaction. MMPs process cytokines, chemokines and growth factors and interfere with chemokine–glycosaminoglycan interaction. Matrix fragments induce chemotaxis of neutrophils, monocytes and fibroblasts and upregulate inflammatory gene expression in macrophages and endothelial cells. Additionally, platelets and plasma proteins participate in the formation of a fibrin-based provisional matrix, which serves as a temporary scaffold for cell migration and modulator of cell behavior [2, 18].

9.3.2   The Proliferative Phase

The proliferative phase is characterized by active ECM deposition and angiogenesis [18]. The secretion of anti-inflammatory mediators [e.g., interleukin-10 (IL-10), transforming growth factor-β (TGF-β)] marks the onset of this phase and is crucial for spatiotemporal confinement of the inflammatory response [2, 21]. The production of TGF-β has a dual effect. Firstly, it plays an important role in fibroblast transdifferentiation into myofibroblasts [2, 21]. Secondly, together with IL-10, it induces the expression of tissue inhibitor of metalloproteinases (TIMP), which is essential for ECM preservation and deposition [18].

Activated myofibroblasts contain stress fibers and α-smooth muscle actin (α-SMA) [21], thus helping them to actively contract and maintain the structural integrity of the infarcted area [18, 23]. They are also highly mitotically active [2] and regulate scar formation through the active synthesis of collagen molecules [18].

Collagen type I and III are deposited as the structural reinforcement of the infarcted region [14, 24] and constitute the predominant components of the scar [25]. Nevertheless, other collagen types are produced as well. Notably, collagen type IV and VI are synthesized in both the infarcted and non-infarcted area, playing a role in cell activation, and the organization of fibrillar collagens and scar [14].

Fibronectin, also present post-MI, helps in fibroblast trans-differentiation into myofibroblasts. Notably, the proliferative phase also triggers the production of matricellular proteins (thrombospondin, tenascin, osteonectin, osteopontin and periostin). Although they do not have any structural role, they are important regulators of cellular behavior including cell adhesion, suppression of inflammation, proper healing, matrix organization and cardiac remodeling [18].

Angiogenesis is stimulated during this phase by vascular endothelial growth factor (VEGF), leading at first to the formation of highly permeable vessels that further support inflammatory cell infiltration [21]. These vessels later mature under the influence of platelet-derived growth factor (PDGF) and thus begin to limit cell extravasation. At the end of this phase, a new vessel network is formed supplying the healing region with oxygen and nutrients [18].

9.3.3   The Maturation Phase

With the accumulation of collagen being synthesized by myofibroblasts, the healing process transitions to the maturation phase. In order to provide sufficient strength to the infarcted region and prevent perforation, the collagen fibers have to be cross-linked by lysyl-oxidases [18]. Myofibroblasts are replaced in the tissue by another differentiated state of fibroblast called the matrifibrocyte, which appears to be more specialized for the lesser metabolically active scar environment [23]. However, this process is far from being complete.

9.4   Post-infarct Remodeling

Following scar formation, if healing has not sufficiently restored the heart's contractile function, the ventricle will continue to change its shape, wall thickness and its mechanical properties [2]. These complex transformations result from the ongoing intraventricular pressure and volume load combined with qualitatively inferior properties of the scar region when compared to the healthy myocardium. These changes are called adverse ventricular remodeling and typically comprise the following: the ventricle becomes more spherical and dilated, and the scar region gets thinner, while the remaining healthy myocardium becomes hypertrophic. Additionally, the infarct area usually expands with time independently of any other MI events, thus slowly worsening the heart condition. Thus, this adverse ventricular remodeling also leads to systolic dysfunction.

Adverse ventricular remodeling correlates with the risk of heart failure development and is associated with chronic inflammation [2, 21]. This suggests possible new targets for the prevention of adverse ventricular remodeling: proper resolving of inflammation [21], confined but mechanically adequate scar [2], and MMPs and matricellular proteins [18].

9.5   Biomaterials for Treatment of Myocardial Infarction

Regenerative medicine has emerged to address the need for cell, tissue and organ regeneration. It implements various approaches, including tissue engineering, stem cell transplantation, small molecules and/or gene therapy [26]. In the past decades, cardiac regeneration has focused on cell therapy, delivery of growth factors, small molecular drugs or use of biomaterials [27–30].

Since the ECM provides essential cues that influence cell phenotype, differentiation, survival and overall tissue homeostasis [18, 31], its protection and/or replacement after MI may simultaneously prevent adverse remodeling and provide a suitable environment to support cell and tissue repair, as well as functional restoration. Biomaterials are an attractive approach for achieving this goal, and various biomaterials are being investigated either as stand-alone therapies or as delivery vehicles for cells, drugs [29, 30] growth factors [32–35], miRNA [36–38], siRNA [39], DNA [40], TIMP [41], oxygen [42], and others [43].

Biomaterials can be categorized according to various parameters. Concerning the chemistry, biomaterials can be of natural or synthetic origin, or a combination of the two [44, 45]. They can be fabricated by various techniques including, but not limited to, electrospinning, phase separation, decellularization, freeze drying [46] 3D printing [47] or self-assembly [48].

Having the inherent excitability of the cardiac tissue in mind, biomaterials for cardiac regeneration can be designed to be electrically non-conductive or conductive [49]. Conductive materials can enhance the functional coupling of healthy and infarcted area and thus improve the ability to contract efficiently. Different means of increasing the electroconductivity have been tested, such as gold nanoparticles, nanowires, carbon nanotubes, graphene oxide, aniline, melanin or polypyrrole [49–61].

There are basically two strategies for delivering biomaterials to the infarcted area: intramyocardial injection or suturing a patch over the afflicted area (Fig. 9.2) [62–64]. Injection is minimally invasive, however, may fail to immediately provide physical stability to prevent left ventricle dilation, depending on the application and type of material used. Patches are more invasive, but can serve as instantaneous mechanical support [62, 65]. The choice of the injection site (infarct region vs. border zone) and the biomaterial spread in the interstitium can affect the therapeutic outcome [66]. Injection into the infarct zone increases cell recruitment to the area as well as mechanical resilience of the wall. Injection into the border zone may salvage more cells and prevent expansion of the infarct. The degree of biomaterial integration within the tissue can affect the electrophysiology of the region.

Both injectable materials and patches can be formulated as hydrogels or non-hydrogel/solid-like materials [64, 65, 67–69]. Hydrogels are highly hydrated matrices with water content $\geq 30\%$ measured by weight [70]. They contain hydrophilic polymers and are able to retain a high water content, yet concurrently they do not dissolve in water or biological fluids due to the presence of cross-links within their

Heart   **Treatment delivery options**

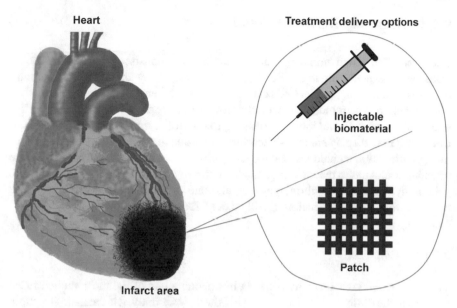

Infarct area

Fig. 9.2 Schematic depicting the most common options for biomaterial delivery

structure [71]. In contrast, non-hydrogel/solid-like materials do not meet these criteria, as they cannot sequester significant amounts of water within their structure.

Hydrogels can be prepared from substances of natural or synthetic origin [70–72]. They can be formed under relatively mild conditions [70], through various chemical and physical methods, which may require external initiation [71–73]. Covalently cross-linked hydrogels are generally more mechanically resistant than physically cross-linked ones [73]. However, physically cross-linked ones offer additional benefits. For example, shear forces can disrupt the cross-links, thus allowing hydrogels that are physically cross-linked to flow easily through narrow spaces (like a syringe), which is called shear-thinning behavior. Once the shear forces stop, the physical cross-links autonomously reform. This self-healing mechanism then recreates the hydrogel. This shear-thinning behavior eliminates the risks associated with premature gelation of covalently cross-linked or thermo-responsive gels, or with diffusion from the site in case of in situ cross-linking [73, 74]. Various shear-thinning and self-healing hydrogels have been designed so far [73, 75–77].

Additionally, novel therapies for MI can be investigated using different in vitro, ex vivo and in vivo models. Recently, the "*Guidelines for experimental models of myocardial ischemia and infarction*" have been published to provide a basic guide for the scientific community in testing cardiac therapies and to ensure better reproducibility and validity of the published data [78].

The following sections outline the various biomaterial therapy strategies and the types of materials being tested.

9.5.1 Naturally Derived Materials

These materials are of animal or plant origin, and they consist of proteins or saccharides naturally occurring in the ECM [44, 45, 79, 80]. They are therefore inherently bioactive and biocompatible. Cells can also trigger their enzymatic degradation, thus allowing them to be remodeled naturally. However, their physicochemical properties and degradation rates are more challenging to control (compared to synthetic materials), they may display batch-to-batch variability due to more inconsistent manufacturing/purification techniques, and they may also be associated with a risk of disease transmission. Among the most investigated naturally derived materials are collagen, gelatin, hyaluronic acid, fibrin, chitosan, alginate or decellularized matrices. Their use for cardiac tissue regeneration will be described in the following paragraphs.

9.5.1.1 Collagen

There are several reasons why collagen has become one of the most studied ECM proteins in the design of biomaterials for MI therapies. Firstly, it is the main structural component of the cardiac ECM [12]. Secondly, collagen interaction with cells mediated by integrin receptors can play a significant role in the control of cell behavior [15, 16]. Thirdly, collagen is rapidly degraded following MI by MMPs into smaller fragments that can regulate inflammation [18]. As for its production, collagen can be either extracted from various tissues of animal origin or biomanufactured as recombinant human collagen in various organisms including plants [81]. Although recombinant collagens offer lesser immunogenic response and reduced risk of disease transmission, current manufacturing technologies are not yet cost-effective, making their use very expensive. Moreover, post-translational modifications of collagen can pose a challenge for biomanufacturing technologies.

To date, the main research focus has been on collagen type I, which has been used to develop both injectable materials and patches. Studies with intramyocardial injections of a rat tail collagen type I hydrogel into the infarct and border zone demonstrated the importance of injection timing [82]. Specifically, earlier treatment post-MI with this material resulted in superior cardiac function and less adverse remodeling. This highlights that different materials may have different optimal windows of opportunity for repairing the post-MI heart. Rat tail collagen type I was also used for making cardiac patches [83]. In order to improve its mechanical properties, dense and mechanically more endurable collagen patches were produced by plastic compression of collagen hydrogel. Patch implantation resulted in neo-angiogenesis, less scar formation and overall preserved cardiac function in a mouse MI model. A commercially available bovine collagen mix of 95% type I and 5% type III improved contractility and prevented paradoxical bulging in infarcted rats [84]. Electrospun calf collagen type I scaffolds were shown to promote cardiomyogenic differentiation

[85], and commercially available collagen constructs enabled the formation of vascularized cardiac patches in vitro [86]. These achievements highlight the potency of collagen as a stand-alone therapy.

Collagen biomaterials have also been used for cell delivery. For example, the injection of circulating angiogenic cells (CACs) within a rat tail collagen type I hydrogel increased CAC engraftment and cardiac function in a mouse MI model [87], and reduced myocardial hibernation and increased wall movement in a swine model [88].

Moreover, various growth factors have been incorporated into collagen-based scaffolds to enhance the healing process. Collagen scaffolds have been designed with growth factor-containing alginate microparticles that gradually release hepatocyte growth factor (HGF) and insulin-like growth factor-1 (IGF-1), which in turn promoted cell proliferation in vitro [35]. Collagen patches with immobilized platelet-derived growth factor (PDGF) were seeded with mesenchymal stem cells and showed superior in vivo effects on cardiac function and vascularization over control patch types in rats [34].

Collagen can also be combined with other substances to fine-tune the biomaterial properties (like stability, electroconductivity or mechanical properties) according to the desired goal. To tailor the mechanical properties while retaining the bioactivity of collagen, Xu et al. developed hybrid injectable hydrogels based on thiolated collagen and oligo(acryloyl carbonate)–poly(ethylene glycol)–oligo(acryloyl carbonate) copolymer (abbreviated Col-SH and OAC–PEG–OAC hydrogel) [89]. These tunable hydrogels were injected either with or without bone-marrow mesenchymal stem cells, improving cardiac function in a rat MI model. In another study, Xia et al. combined collagen type I with N-isopropylacrylamide/acrylic acid/2-hydroxyethyl methacrylate-poly-ε-caprolactone [90]. Injection of this thermosensitive hydrogel with mesenchymal stem cells improved cell engraftment and functional outcome in a mouse MI model. Electrospun nanofiber patches based on collagen type I, elastin and polycaprolactone combined the bioactivity of different natural substances with the superior mechanical properties of a synthetic polymer [91]. Upon implantation, these patches showed good outcomes both with and without c-kit$^+$ cells in a mouse MI model. Shafiq et al. [92] incorporated bioactive substances, which could be slowly released, into electrospun patches from collagen type I and polycaprolactone. Patches containing substance P alone and in combination with insulin-like growth factor-1C (IGF-1C) peptide promoted cardiac function and angiogenesis in vivo. In another study, Reis et al. [93] incorporated a pro-survival angiopoietin-1-derived peptide (QHREDGS) into a chitosan-collagen hydrogel. Treatment of MI rat hearts using this peptide-modified hydrogel led to improved morphology and function of the myocardium.

To render the biomaterial electroconductive, the incorporation of nanosilver and nanogold particles into rat tail collagen type I hydrogels has been investigated [94]. These electroconductive matrices improved proliferation and function of neonatal rat ventricular cardiomyocytes under electrical stimulation. Similar hybrid cardiac patches based on a porcine collagen type I hydrogel and nanogold-containing electrospun collagen fibers led to increased expression of connexin-43 (Cx43) and improved cardiac function in a mouse MI model [50]. The incorporation of single-walled carbon nanotubes is another approach for enhancing electrical coupling between cells. For example, single-walled carbon nanotubes were combined with collagen [52] or with collagen and chitosan [53] and were able to support cell growth and beating function. Another option to confer electroconductive properties is to introduce graphene oxide [49]. In one study, collagen type I scaffolds were fabricated by freeze drying, then covalently coated with graphene oxide, and reduced to restore graphene oxide electroconductivity. These scaffolds promoted the expression of cardiac genes, including the *Cx43* gene involved in electrical coupling.

Seeing the potential of collagen, it is not surprising that a collagen type I material has entered a clinical trial. The MAGNUM trial evaluated the efficacy of a coronary artery bypass graft combined with the implantation of a collagen type I scaffold and autologous mononuclear bone-marrow cells [95]. Although cardiac function was improved, it could not be conclusively attributed to cell-scaffold therapy only. Additional trials will be needed to unravel the effect of collagen as a stand-alone treatment.

Although the main research focus has been on collagen type I, to a lesser extent collagen type III has been investigated as well. Injectable collagen/glycol-chitosan hybrid hydrogels containing collagen type I, type III, or both type I and III were prepared from human tropocollagen molecules [96]. Tropocollagen was fibrillated during the fabrication of the hydrogel. The resulting scaffold was highly porous and promoted cell adhesion and migration. A commercially available bovine collagen mix of 95% type I and 5% type III [84] has also been under investigation, as was already discussed above.

Altogether, collagen biomaterials offer tremendous potential for repairing the infarcted heart. Although currently expensive, recombinant human collagen would be a good candidate for future studies in order to overcome the challenges of products of animal origin. Additional clinical studies are also needed to translate collagen into a stand-alone therapy.

9.5.1.2 Gelatin

Gelatin is another option for cardiac regeneration. It is derived from collagen, is biodegradable and has low immunogenicity and also a low cost [97]. It has long been used in the food, cosmetic and pharmaceutical industries.

Gelatin hydrogels were shown to increase the engraftment of fetal rat cardiomyocytes as well as to enhance angiogenesis and cardiac function in a rat MI model [98]. Gelatin materials have also been used to deliver growth factors, cytokines, genes or miRNA to promote MI healing. For example, basic fibroblast growth factor (bFGF, also known as FGF-2) was incorporated into gelatin hydrogels to enhance angiogenesis in a rat [32] and canine model [33], which led to improved cardiac function. Its use in delivering a secretome product, i.e., a mixture of bioactive factors (cytokines, growth factors and exosomes) produced by cells was also investigated [43]. The secretome produced by human adipose-derived stem cells was incorporated into an injectable shear-thinning hydrogel made of gelatin and Laponite® (nanoclay). Peri-infarct delivery resulted in increased angiogenesis and reduced cardiac remodeling in rats. An injectable gelatin–silicate hydrogel was used to deliver miR-1825 into the mouse heart post-MI [36], which induced cardiomyocyte proliferation and improved cardiac function. Polyethylenimine-functionalized graphene oxide nanosheets complexed with vascular endothelial growth factor-165 DNA were incorporated in a methacrylated gelatin hydrogel to function as a non-viral gene delivery system; its use led to increased angiogenesis and reduced scar size after injection in rats [40].

Like collagen, gelatin has been combined with other substances to increase its electroconductivity [99, 100], or tailor its mechanical properties [97]. Electrospun gelatin and poly-ε-caprolactone patches loaded with mesenchymal stem cells enhanced cell survival and improved cardiac function in a rat MI model [65]. Noshadi et al. [101] developed gelatin methacrylate (GelMA) hydrogels that were injected into the infarct site and then photo-cross-linked in situ using visible light. GelMA hydrogels were also used with carbon nanotubes [100] or electrically conductive poly(thiophene-3-acetic acid) [102], which enhanced the electrical coupling of cells. Additionally, electroconductive nanoparticles based on GelMA and polypyrrole were cross-linked onto electrospun GelMA-polycaprolactone nanofibrous membranes to form electroconductive cardiac patches [55]. These then promoted cardiac function, reduced infarct size and increased angiogenesis in vivo. Lastly, single-walled carbon nanotubes combined with pure gelatin into an electroconductive hydrogel also improved cardiac function in rats [99].

In addition to being used as the main component of a given material, gelatin can also be used for surface modification. Wang et al. used an iron oxide framework and coated it with gelatin. Thereafter, bone-marrow-derived rat mesenchymal stem cells were seeded onto the coated framework and allowed to proliferate [103]. Prior to patch transplantation into rats, cell-seeded frameworks were coated again, this time with Matrigel. This cell-seeded patch implantation resulted in improved cardiac function.

In summary, gelatin offers high biocompatibility, can be easily combined with other substances, and can additionally be used as a coating, making it an attractive candidate for cardiac regeneration purposes.

9.5.1.3 Hyaluronic Acid

The glycosaminoglycan hyaluronic acid is another natural component of the heart [11] used in engineering cardiac materials. Hyaluronic acid is a polysaccharide composed of repeating N-acetylglucosamine and glucuronic acid units, which can be cross-linked into hydrogels and easily modified [45]. However, it lacks mechanical strength. A simple hyaluronic acid hydrogel loaded with human cord blood mononuclear cells showed improved cardiac function in a pig model and was superior to the injection of cells or hydrogel only [104].

A commercially available hyaluronic acid-based product Extracel-HP™ was combined with platelet-rich plasma to deliver growth factors and cytokines involved in healing and angiogenesis [105]. Extracel-HP™ consists of thiol-modified hyaluronate, heparin, gelatin, polyethylene glycol and degassed water. When combined with platelet-rich plasma, ascorbic acid, ibuprofen and allopurinol and then injected into pigs post-MI, it showed a superior therapeutic effect. Hyaluronic acid-based shear-thinning and self-healing hydrogels were tested in rats and sheep [74, 106] and have been combined with miR-302 [37], siRNA [107] and extracellular vesicles [108]. MiR-302 mimics were able to promote cardiomyocyte proliferation in vivo, thus targeting enhancement of cardiac regeneration. siRNA delivery focused on adverse remodeling post-MI, as it interfered with mRNA for MMP2 and therefore specifically attenuated MMP2 expression after MI. Extracellular vesicles are interesting due to their pro-angiogenic, pro-proliferative and anti-apoptotic effects via paracrine mechanisms [108]. Chen et al. derived extracellular vesicles from endothelial progenitor cells, embedded them into a shear-thinning hydrogel and observed enhanced angiogenesis and cardiac function in a rat MI model. Delivery of miRNA was performed also using a commercially available Glycosan HyStem® hyaluronic acid-based hydrogel [38]. As the miR-29 family targets genes associated with ECM (collagen, elastin, MMP2, laminin and others), miR-29B delivery to the border zone was shown to influence ECM remodeling after MI in mice. Direct inhibition of MMPs was investigated as well in a study in which Purcell et al. delivered recombinant TIMP-3 using a hyaluronic acid-based hydrogel [41]. They incorporated a specific peptide sequence (GGRMSMPV) into the hydrogel to render the matrix sensitive to intrinsic MMP cleavage, thus tailoring the release kinetics of TIMP-3 from the hydrogel. Injection of this TIMP-3 releasing hydrogel led to attenuation of adverse remodeling in a pig model. Fan et al. [109] targeted MMP-2 and MMP-9, both upregulated after MI, using the specific peptide inhibitor CTTHWGFTLC. Delivery of this peptide reduced ECM degradation and led to improved cardiac function in rats. Wang et al. [110] delivered nanoparticles containing plasmid DNA encoding endothelial nitric oxide synthase (eNOs) together with adipose-derived stem cells within an electroconductive hyaluronic acid-based hydrogel. This led to increased cardiac function, greater vessel density and less scarring.

In a combined approach, hyaluronic acid was mixed with hydroxyethyl methacrylate [111]. This hydrogel improved cardiac function and increased infarct thickness and stiffness in a porcine model. All these results confirm the suitability of hyaluronic acid for MI treatment and the possibility of combining it with various bioactive molecules.

9.5.1.4 Fibrin

Fibrin is another intuitive choice for cardiac biomaterial therapy, as it is present at the very beginning and throughout much of the MI healing process, in the form of a fibrin-based provisional matrix [2, 18]. For example, an easily applicable fibrin cardiac patch was established using a fibrin gel spraying system capable of in situ polymerization [112]. The setup efficiency was successfully verified in a mouse model and holds promise due to its ease of use.

Like most materials being developed, fibrin patches have also been used to deliver various cell types in numerous animal models. For example, mouse embryonic stem cells [113], human ESC-derived endothelial cells and smooth muscle cells [114], human ESC-derived cardiac progenitors [115], human bone-marrow-derived mesenchymal stem cell [116], human induced pluripotent stem cell (iPSC)-derived cardiomyocytes and human pericytes [117], human iPSC-derived cardiomyocyte spheroids [118] and many others have been tested, as reviewed elsewhere [64, 119, 120]. Of note is that a fibrin scaffold with human ESC-derived cardiac progenitors has been tested in humans [121]. A 68-year-old patient suffering from severe heart failure had this cell-seeded scaffold implanted onto the infarct area. An increase in cardiac function together with abatement of the symptoms was observed three months after surgery.

To further promote the clinical applicability of fibrin, large fibrin patches containing human iPSC-derived cardiomyocytes, smooth muscle cells and endothelial cells were developed [122]. These patches were first cultured under dynamic conditions, which resulted in spontaneous contraction and patch maturation. They were subsequently evaluated in a porcine MI model where they improved cardiac function and reduced infarct size. These results confirm the feasibility of large patch formation. However, as oxygen and nutrient delivery to the infarcted area remains a challenge, microvessel incorporation into the patch could significantly improve the outcomes after MI. Therefore, Riemenschneider et al. [123] evaluated fibrin patches containing microvessels in a rat MI model. The authors used their previous observations that microvessels self-assembled from blood outgrowth endothelial cells and pericytes in a fibrin gel after 5 days of in vitro culture [124]. Fibrin patches implanted into rats contained either randomly oriented or aligned microvessels, a portion of which were successfully perfused after implantation. Su et al. [125] developed a vascularized patch with the use of microfluidic technology. First, microvessels with human umbilical vein endothelial cells were formed by hydrodynamic focusing in microfluidic channels followed by photopolymerization. Second, after cultivation, these microvessels were arranged and together with cardiac stem cells embedded into a fibrin gel. These constructs were subsequently transplanted into rats post-MI, leading to increases in cell proliferation and angiogenesis in the peri-infarct area. 3D printing is another high-tech approach to engineer cardiac tissue in vitro. Wang et al. mixed primary cardiomyocytes with fibrin-based bio-ink and printed a tissue construct, which upon maturation, contracted and expressed markers typical for cardiac tissue [126].

Growth factor delivery has also been investigated with promising outcomes. Polyethylene glycol-fibrinogen hydrogels were used to deliver VEGF and angiopoietin-1 in a rat model [127]. Bearzi et al. [128] generated iPSC-derived cardiomyocytes engineered to secrete placental growth factor and MMP9. Mice injected with these transduced cardiomyocytes embedded within a polyethylene glycol–fibrinogen scaffold showed improved cardiac functions and angiogenesis.

Given that fibrin has shown very good outcomes for MI treatment and that it is currently already approved for clinical use as hemostat, sealant, and adhesive by the Food and Drug Administration (FDA) [129], the path to its transition to a real clinical therapy for cardiac regeneration seems very promising.

9.5.1.5 Decellularized Matrix

A natural ECM for therapy can be obtained by the decellularization of native tissue or even of the whole organ [130]. In this process, cells are gently removed yielding a natural ECM scaffold with preserved architecture. This natural ECM can be subsequently used either as a stand-alone therapy or be secondarily re-cellularized by cells of interest. Various types of tissues have been decellularized and evaluated in MI models, altogether highlighting the importance of ECM-mediated signaling.

As a stand-alone therapy, Wassenaar et al. injected decellularized and partially digested porcine ventricular myocardial matrix into rat hearts post-MI [131]. This hydrogel matrix injection activated numerous repair responses in the myocardium, such as enhanced angiogenesis and progenitor cell recruitment, reduced apoptosis, hypertrophy and fibrotic response, and positive changes in cardiac function. Similar results were obtained by Sarig et al. who used a decellularized porcine cardiac ECM patch implanted onto the rat heart either in an acute or chronic MI model [132].

Decellularized ECM from the heart has also been used to enhance cell delivery and retention. Adipose-derived stem cell delivery using a decellularized porcine myocardial ECM patch [133] and hiPSC-derived cardiac cell delivery with decellularized rat cardiac ECM [134] were both shown to improve function in rat MI models. Mesenchymal stem cell-seeded decellularized rabbit pericardial patches [135, 136], or human heart valve-derived patches either alone or combined with bone-marrow c-kit$^+$ cells [137] also led to improved cardiac function. Decellularized ECM from non-cardiac tissue types has been tested for heart repair as well. Decellularized umbilical arteries from humans were used as scaffolds to deliver human mesenchymal stem cells in a rat MI model [138], and an injectable matrix from decellularized porcine small intestinal submucosa was investigated with or without circulating angiogenic cells (CACs) in a mouse MI model [139].

Decellularized ECM loaded with growth factors has also been used successfully for treating the MI heart. For example, a decellularized porcine pericardium containing HGF led to a reduction of adverse remodeling and improved cardiac function in vivo [140]. A hydrogel derived from decellularized human placenta containing ECM components and growth factors also showed a pro-healing effect in rats [141]. To develop an even more intricate treatment, Wang et al. took advantage of 3D dynamic culture conditions [39]. They mixed bone-marrow mesenchymal stem cells with RAD16-I peptide hydrogel (PuraMatrix®) and added that onto decellularized porcine pericardium with bound vascular endothelial growth factor (VEGF). Following cultivation under dynamic conditions in a perfusion system, these cell–ECM sheets were implanted into rats post-MI and showed better improvement than both control cell–ECM sheets cultured under static conditions and the decellularized matrix only. Another pre-vascularized stem cell- and VEGF-containing patch was prepared using 3D printing technology [47]. Two types of bio-ink were printed in a specific pattern: one contained cardiac progenitor cells, the other mesenchymal stem cells and VEGF. These patches showed high vascularization and enhanced cardiac function when transplanted into mice.

Two commercially available products have been evaluated in clinical trials. After showing positive effects in porcine, rat and mouse MI models [142–144], as well as in calves with chronic ischemic heart failure [145], small intestinal submucosa CorMatrix® was evaluated in a clinical study for use in humans (NCT02887768) [143, 146]. Secondly, an injectable hydrogel from decellularized porcine myocardial ECM [147], and commercially available as VentriGel™, was investigated in rats and pigs [148, 149], and in a phase I clinical trial (NCT02305602) [150, 151].

In summary, the natural bioactivity and architecture of decellularized ECM makes it a very promising candidate for MI treatment.

9.5.1.6 Alginate

Alginate is polysaccharide isolated from seaweed [45]. It has been reported to be biocompatible, but not degradable in the human body, and it can form gels. For MI treatment, an alginate hydrogel as a cell delivery vehicle was shown to increase retention of mesenchymal stem cells injected into the heart in a pig model [152]. Alginate hydrogels have also been tested for the delivery of growth factors and genes. The combination of alginate with growth hormone in a rat MI model showed enhanced cardiac function, angiogenesis, activation of myofibroblasts and electrical conduction [153, 154]. The delivery of 6-bromoindirubin-3-oxime and insulin-like growth factor 1 (IGF-1) using gelatin nanoparticles and an alginate hydrogel promoted cardiomyocyte proliferation and angiogenesis in rats [155]. VEGF delivery using alginate microspheres in a chitosan patch was found to promote angiogenesis in MI rats [156]. As for gene delivery, synthetic modified mRNA termed M3RNA (microencapsulated modified messenger RNA) was embedded into alginate and upon injection into the pig heart achieved rapid induction of protein expression [157]. This is a good proof of concept, as gene delivery therapies can be hindered by inadequate activation of gene expression.

In order to promote electroconductivity, Dvir et al. combined an alginate scaffold with gold nanowires and showed better contraction and coupling of neonatal rat cardiomyocytes cultured on the scaffolds in vitro [51].

Regarding clinical translation, Leor et al. developed an injectable alginate-gluconate-based scaffold that had promising results in a porcine MI model [158]. This scaffold became commercially available under the name IK-5001 and was tested in a small clinical trial, where it preserved left ventricular ejection fraction [159]. These results led to a large randomized, double-blind, placebo-controlled trial—the PRESERVATION I trial [160]. However, the clinical data did not confirm efficacy for IK-5001 administration [151, 161]. Another commercially available alginate-based hydrogel was tested in larger animals and clinical trials for its ability to improve the outcomes in the failing heart. Specifically, the 2-component alginate self-gel formulation hydrogel (Algisyl-LVR™) was injected into hearts of dogs with chronic heart failure via open chest surgery, which significantly improved hemodynamic parameters of the left ventricle [162]. A small human clinical trial then followed combining Algisyl-LVR™ injection with coronary artery bypass grafting in patients with failing heart [163]. This study showed improved cardiac function and restoration of ventricular geometry, thus confirming the promising results from animal studies. A multicenter randomized controlled clinical trial (AUGMENT-HF) subsequently compared the efficacy of standard medical therapy versus standard medical therapy combined with Algisyl-LVR™ injection in patients with severe heart failure [164]. This randomized trial demonstrated a benefit of Algisyl-LVR™ injection in long term, highlighting the promise of this therapy for implementation in the clinic.

9.5.1.7 Chitosan

Chitosan is a natural polymer composed of D-glucosamine and N-acetyl-glucosamine [45]. It is prepared by partial deacetylation of chitin, a component of arthropod exoskeletons. On its own, a chitosan hydrogel showed increased cardiac function and reduced infarct size when used to treat rats with MI [165]. It was also used to enhance the retention and therapeutic effect of mesenchymal stem cells in a rat MI model [166].

However, chitosan has been mainly investigated in combination with other substances. The synthetic peptide RoY (YPHIDSLGHWRR) was previously shown to enhance angiogenesis under hypoxic conditions both in vitro and in vivo [167]. Therefore, its combination with a chitosan hydrogel was evaluated in a rat MI model, where it improved angiogenesis and cardiac function [168]. Electrospun cardiac patches from cellulose, chitosan and silk fibroin combined with adipose-derived mesenchymal stem cells [169], and aligned electrospun chitosan/calcium silicate nanofiber patches seeded with neonatal rat cardiomyocytes [170] have also both been shown to improve cardiac function in rat MI models. In other work, a conductive hydrogel was obtained by combining chitosan and polypyrrole [58, 171]. The gels improved the electrical conduction as well as cardiac function upon injection into the infarcted rat heart. Another design used chitosan and graphene oxide-gold nanosheets to increase scaffold electroconductivity [59]. In a completely different approach, an auxetic conductive cardiac patch was developed by Kapnisi et al. [60] (Fig. 9.3). Unlike the majority of other materials, auxetic materials can expand simultaneously in multiple directions when stretched, which offers extraordinary benefits when applied as a cardiac patch. Briefly, laser microablation was used to introduce a re-entrant honeycomb (bow-tie) micropattern to chitosan films. These were then coated with polyaniline and phytic acid, in which the polyaniline provided electroconductivity. This patch with auxetic design and electroconductive properties integrated well in vivo in a rat MI model and led to reduced LV mass (i.e., less hypertrophy) compared to the control group. Thus, auxetic materials hold promise for future investigation.

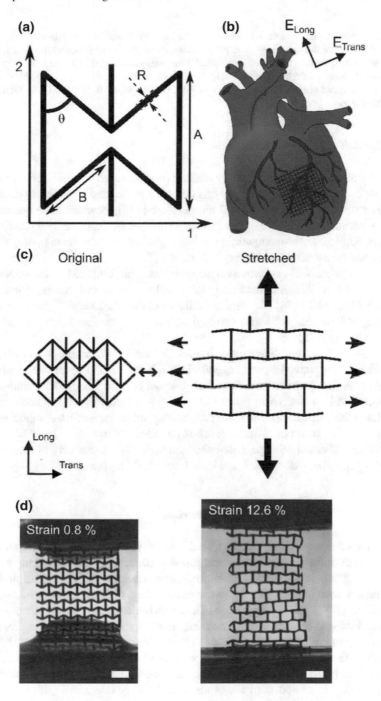

◄**Fig. 9.3** Auxetic cardiac patch. **a** Schematic of the bow-tie dimensions. **b** Schematic illustration of the alignment of the auxetic cardiac patch (AuxCP) on the heart. **c** Schematic illustration of the auxetic behavior of the re-entrant honeycomb (bow-tie) geometry. **d** Digital optical microscope images of the AuxCPs during tensile testing at 0.8% strain and 12.6% strain (scale bars: 1 mm). Taken from Kapnisi et al. [60]. Copyright Wiley-VCH Verlag GmbH & Co. KGaA. Reproduced with permission

9.5.1.8 Self-assembling Hydrogels

Self-assembly is a spontaneous and reversible process based on weak (non-covalent) interactions [48, 172]. The assembly can be finely modulated by various parameters such as molecular structure, pH and temperature [173]. Peptides have been recognized as particularly suitable molecules for self-assembly, especially those forming β-sheets. Peptides are biocompatible and biodegradable and can form hydrogels with nanoarchitecture resembling the natural ECM [172].

In one example, self-assembling hydrogels based on folic acid and peptides were combined with iPSCs and showed greater cell retention and cardiac function in a mouse MI model [174]. A commercially available PuraMatrix® was used with mesenchymal stromal cells and resulted in enhanced cardiac function and MSC retention in rats [175].

Moreover, self-assembling peptide hydrogels were also functionalized with a peptide mimic of the Notch1 ligand Jagged1 [176]. Notch ligand-containing hydrogels improved cardiac function and decreased fibrosis in a rat MI model. PuraMatrix® was also used in a biohybrid patch [177]. A porous poly(ethyl acrylate) scaffold was filled with adipose-derived stem cells entrapped in the self-assembling peptide PuraMatrix® gel to produce large patches intended for use in a sheep MI model. Many other self-assembling peptide-based strategies for the delivery of therapeutics to treat MI have been developed and have been nicely reviewed [48].

9.5.1.9 Other Naturally Derived Materials

Among the other natural substances tested for cardiac regeneration are, for example, sericin and albumin. Sericin is a natural protein produced by the silkworm as part of its cocoon. It is biocompatible and antioxidative, enhances adhesion and proliferation, and has been investigated for tissue engineering, pharmaceutical and cosmetic applications [178, 179]. An injectable silk sericin hydrogel was tested in a mouse MI model and showed pro-angiogenic, anti-inflammatory and anti-apoptotic properties, leading to improved cardiac function [179]. In another study, electrospun albumin scaffolds with adsorbed gold nanorods were seeded with cardiac cells to create a cardiac patch [180]. These patches were attached to the rat heart via a suture-free approach using laser and gold nanorods, and it nicely integrated with the cardiac tissue.

In summary, there are many natural substances that have shown great promise for the treatment of myocardial infarction, some of which have proceeded to clinical trials.

9.5.2 Synthetic Materials

These materials are artificially synthesized and can have numerous different components and chemical compositions [44, 45, 79, 80]. Synthetic materials have tailorable physicochemical properties and degradation rate, can be manufactured on a large scale, have minimal batch-to-batch variability, and can be processed into scaffolds using various techniques. However, although synthetic materials used in tissue engineering can be made biocompatible, they typically lack inherent bioactivity. Consequently, these materials are often subjected to various physical or chemical modifications in order to introduce functional groups for cell recognition and binding. Polyethylene glycol, polycaprolactone, polyvinyl alcohol and other synthetic materials have been tested for cardiac tissue regeneration, and they will be briefly introduced in the following sections.

9.5.2.1 Polyethylene Glycol

Polyethylene glycol (PEG) is synthetic, biocompatible, hydrophilic, water-soluble and FDA-approved polymer [45, 181]. Injectable PEG hydrogels were tested in rats in order to evaluate the influence of hydrogel interstitial spreading on electrophysiological properties of the heart [66]. No conduction abnormalities were observed in hearts with highly spread hydrogels. However, hearts with a minimal spread of the hydrogel (i.e., bolus injections) were prone to arrhythmias. This work highlights the importance of considering the delivery site and gel spread properties.

Combined PEG material with cell and growth factor delivery has been investigated as well. A PEG hydrogel loaded with hiPSC-derived cardiomyocytes and erythropoietin improved cardiac function and attenuated remodeling in a rat model [182]. However, to date, PEG has mostly been used in combination with other components to produce hybrid materials. Dong et al. [56] developed a self-healing, conductive injectable hydrogels based on chitosan, aniline and dibenzaldehyde-terminated polyethylene glycol for cell delivery. In another study, polyethylene glycol-dimethacrylate hydrogel patches with encapsulated stem cells were glued using fibrin glue onto the infarcted area of mouse hearts [183]. An injectable polyethylene glycol-hyaluronic acid-based hydrogel with Wharton's jelly mesenchymal stem cells and insulin-like growth factor-1 promoted the greatest improvement in contractile function and new vessel formation in a rabbit MI model [184]. Steele et al. [185] combined a shear-thinning self-healing hydrogel based on peptide and polyethylene glycol with a dimeric fragment of HGF and observed enhanced cardiac function, angiogenesis and reduced scar in a rat MI model. Ciuffreda et al. [186] introduced heparin into polyethylene glycol (PEG)-based hydrogel with an MMP1 degradable peptide sequence (GCREGPQGIWGQERCG) and combined the heparin-containing hydrogel with bone-marrow-derived mesenchymal stromal cells. They observed better cell engraftment, increased cardiac function and angiogenesis in a rat MI model. Lastly, electroconductive polyethylene glycol-based materials were also developed. Specifically, Bao et al. [57] tested an injectable conductive polyethylene glycol-melamine-hyaluronic acid-based hydrogel containing graphene oxide. When combined with adipose tissue-derived stromal cells, it improved cardiac function and enhanced electrical coupling in rats. Zhou et al. [61] investigated conductive hydrogels based on polyethylene glycol, fumarate and graphene oxide. These hydrogels improved electrophysiological and cardiac function also in rats.

Controllable physical properties and the ability to form gels make PEG a good biomaterial candidate for MI therapies.

9.5.2.2 Polycaprolactone

Poly-ε-caprolactone (PCL) is a synthetic biocompatible polymer with prolonged degradation time, tailorable mechanical properties and suitability for forming blends and co-polymers [45, 187]. It is extensively used for tissue engineering and pharmaceutical applications, and it is FDA-approved [187]. Its major drawback is its hydrophobicity. As a result, various physical and chemical modifications have been implemented to render it more bioactive. For cardiac regeneration purposes, it has been used in combination with other substances.

Soler-Botija et al. [188] developed PCL-methacryloyloxyethyl ester-based scaffolds, which were filled with a self-assembling peptide hydrogel (PuraMatrix®). These scaffolds were seeded with adipose tissue-derived progenitor cells and were found to enhance cardiac function in a mouse MI model. Chung et al. photochemically grafted silk fibroin-PCL-based patches with hyaluronic acid and GRGD (an $\alpha_v\beta_3$ integrin ligand), which promoted greater cardiomyogenic differentiation of bone-marrow-derived mesenchymal stem cells in vitro [189]. In another study, a hydrogel based on PCL, 2-hydroxyethyl methacrylate (HEMA) and poly-N-isopropylacrylamide (PNIPAAm) was used for VEGF growth factor delivery [190]. Injection of this PCL-HEMA/PNIPAAm hydrogel combined with VEGF led to enhanced angiogenesis and improved cardiac function in a rat MI model. The same authors also investigated FGF2 delivery using the same PCL-HEMA/PNIPAAm hydrogel [191]. Results again showed improved angiogenesis and cardiac function in rats.

Outside of these two highly investigated synthetic materials (PEG and PCL), many other synthetic substances have also been tested for cardiac repair, some of which are highlighted in the next section.

9.5.2.3 Other Synthetic Materials

Polyvinyl alcohol (PVA) is a synthetic biocompatible polymer [192]. The injection of a PVA hydrogel was investigated in a sheep model of chronic ischemic mitral regurgitation post-MI, where it reduced the severity of mitral regurgitation as well as ventricular remodeling [192]. Another PVA-based material was developed by Tang et al. in which the PVA was first molded into a unique microneedle patch with porous structure (Fig. 9.4) [193]. Next, cardiac stromal cells were embedded into fibrin gel and placed on the basal part of the microneedle patch. The sophisticated porous structure of the patch then enabled for various factors secreted by cardiac stromal cells to be delivered through the patch and microneedles into the region of patch implantation. The cell-seeded microneedle patch was tested in a rat and porcine MI model and resulted in enhanced cardiac function and prevention of adverse remodeling.

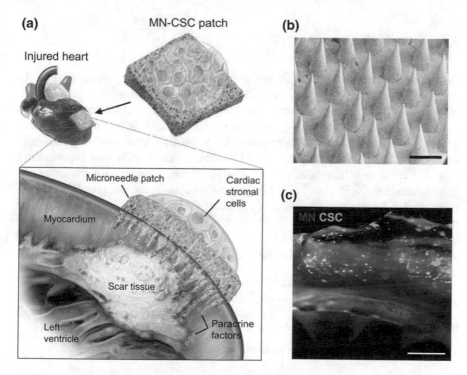

Fig. 9.4 Microneedle patch strategy for MI repair. **a** Schematic showing the overall design used to test the therapeutic benefits of MN-CSCs on infarcted heart. **b** SEM image of MN. Scale bar, 500 μm. **c** Representative fluorescent image indicating that DiO-labeled CSCs (green) were encapsulated in fibrin gel and then integrated onto the top surface of MN array (red). Scale bar, 500 μm. Reprinted with permission of AAAS from Tang et al. [193]. © The Authors, some rights reserved; exclusive licensee American Association for the Advancement of Science. Distributed under a Creative Commons Attribution NonCommercial License 4.0 (CC BY-NC) http://creativecommons.org/licenses/by-nc/4.0/

Examples of other synthetic polymers tested for treatment of MI are polylactic-co-glycolic acid [194], poly-L-lactic acid [195], poly-L-lactide-co-caprolactone and polyethyloxazoline [68], N-isopropylacrylamide (NIPAAm), N-vinylpyrrolidone (VP) and methacrylate-polylactide [196, 197], as well as NIPAAm and hydroxyethyl methacrylate [198].

Two interesting studies on oxygen-releasing scaffolds were published recently. In one, a hydrogel based on NIPAAm, HEMA and acrylate-oligolactide was combined with oxygen release microspheres with a core–shell structure [42]. This injectable system was able to release oxygen, stimulate angiogenesis and improve cardiac function in rats. In the second study, an oxygen-releasing antioxidant polymeric cryo-gel scaffold was based on polyurethane and calcium peroxide [199]. Although tested only in an ischemic flap model, the scaffold may also hold promise for MI treatment, as it was able to efficiently release oxygen and prevent tissue necrosis.

9.6   Conclusions

Bioengineered materials developed to date offer novel and attractive strategies for the treatment of MI. They can provide not only a suitable ECM environment that supports cell survival, growth and tissue repair, but they can also prevent adverse remodeling post-MI. As both natural and synthetic materials have their inherent advantages and challenges, the choice of biomaterial has to be done carefully with respect to desired outcomes. Moreover, natural and synthetic biomaterials may be combined to create even more advanced and tunable scaffolds. Studies have demonstrated that biomaterials can be used either as stand-alone therapies, or be combined with cells, growth factors, small molecules and other therapeutics to further enhance the therapeutic effect. The evidence from the accumulating research is making clinical translation seem imminent, but additional large animal studies and clinical trials will be needed to successfully implement biomaterial therapy in the clinic for the treatment of MI patients.

Acknowledgements This work was supported by a Collaborative Research Grant from the Canadian Institutes of Health Research (CIHR) and the Natural Sciences and Engineering Research Council (NSERC) (CPG-158280), and a CIHR operating grant (MOP-77536).

Disclosure All authors have read and approved the final version.

References

1. WHO. Global health estimates 2016: deaths by cause, age, sex, by country and by region, 2000–2016. Geneva; 2018.
2. Steenbergen C, Frangogiannis NG. Ischemic heart disease, chap. 36. In: Hill JA, Olson EN, editors. Muscle. Boston/Waltham: Academic Press; 2012. p. 495–521.

3. Waller DG, Sampson AP. Ischaemic heart disease, chap. 5. In: Waller DG, Sampson AP, editors. Medical pharmacology and therapeutics, 5th ed. Elsevier; 2018. p. 93–110.
4. WHO. International classification of diseases 11th revision. https://icd.who.int/. 21 Dec 2018.
5. Roth GA, Johnson C, Abajobir A, Abd-Allah F, Abera SF, Abyu G, Ahmed M, Aksut B, Alam T, Alam K, et al. Global, regional, and national burden of cardiovascular diseases for 10 causes, 1990 to 2015. J Am Coll Cardiol. 2017;70(1):1–25.
6. Reed GW, Rossi JE, Cannon CP. Acute myocardial infarction. The Lancet. 2017;389(10065):197–210.
7. Savarese G, Lund LH. Global public health burden of heart failure. Card Fail Rev. 2017;3(1):7–11.
8. Azad N, Lemay G. Management of chronic heart failure in the older population. J Geriatr Cardiol. 2014;11(4):329–37.
9. Bonacchi M, Harmelin G, Bugetti M, Sani G. Mechanical ventricular assistance as destination therapy for end-stage heart failure: has it become a first line therapy? Front Surg 2015;2(35).
10. Laflamme MA, Sebastian MM, Buetow BS. Cardiovascular, chap. 10. In: Treuting PM, Dintzis SM, editors. Comparative anatomy and histology. San Diego: Academic Press; 2012. p. 135–53.
11. Jourdan-LeSaux C, Zhang J, Lindsey ML. Extracellular matrix roles during cardiac repair. Life Sci. 2010;87(13):391–400.
12. Horn MA, Trafford AW. Aging and the cardiac collagen matrix: novel mediators of fibrotic remodelling. J Mol Cell Cardiol. 2016;93:175–85.
13. LeGrice I, Pope A, Smaill B. The architecture of the heart: myocyte organization and the cardiac extracellular matrix. In: Villarreal FJ, editor. Interstitial fibrosis in heart failure. New York, NY: Springer; 2005. p. 3–21.
14. Shamhart PE, Meszaros JG. Non-fibrillar collagens: key mediators of post-infarction cardiac remodeling? J Mol Cell Cardiol. 2010;48(3):530–7.
15. Zeltz C, Gullberg D. The integrin–collagen connection—a glue for tissue repair? J Cell Sci. 2016;129(4):653.
16. Shattil SJ, Kim C, Ginsberg MH. The final steps of integrin activation: the end game. Nat Rev Mol Cell Biol. 2010;11:288.
17. Xu J, Mosher D. Fibronectin and other adhesive glycoproteins. In: Mecham RP, editor. The extracellular matrix: an overview. Berlin, Heidelberg: Springer; 2011. p. 41–75.
18. Dobaczewski M, Gonzalez-Quesada C, Frangogiannis NG. The extracellular matrix as a modulator of the inflammatory and reparative response following myocardial infarction. J Mol Cell Cardiol. 2010;48(3):504–11.
19. Lindsey ML. MMP induction and inhibition in myocardial infarction. Heart Fail Rev. 2004;9(1):7–19.
20. Sternlicht MD, Werb Z. How matrix metalloproteinases regulate cell behavior. Annu Rev Cell Dev Biol. 2001;17(1):463–516.
21. Prabhu SD, Frangogiannis NG. The biological basis for cardiac repair after myocardial infarction: from inflammation to fibrosis. Circ Res. 2016;119(1):91–112.
22. Hodgkinson CP, Bareja A, Gomez JA, Dzau VJ. Emerging concepts in paracrine mechanisms in regenerative cardiovascular medicine and biology. Circ Res. 2016;118(1):95–107.
23. Fu X, Khalil H, Kanisicak O, Boyer JG, Vagnozzi RJ, Maliken BD, Sargent MA, Prasad V, Valiente-Alandi I, Blaxall BC, et al. Specialized fibroblast differentiated states underlie scar formation in the infarcted mouse heart. J Clin Invest. 2018;128(5):2127–43.
24. Wei S, Chow LT, Shum IO, Qin L, Sanderson JE. Left and right ventricular collagen type I/III ratios and remodeling post-myocardial infarction. J Card Fail. 1999;5(2):117–26.
25. Sun Y, Weber KT. Infarct scar: a dynamic tissue. Cardiovasc Res. 2000;46(2):250–6.
26. Mason C, Dunnill P. A brief definition of regenerative medicine. Regen Med. 2007;3(1):1–5.
27. Vadakke-Madathil S, Chaudhry Hina W. Cardiac regeneration. Circ Res. 2018;123(1):24–6.
28. Tzahor E, Poss KD. Cardiac regeneration strategies: staying young at heart. Science. 2017;356(6342):1035.

29. Cambria E, Pasqualini FS, Wolint P, Günter J, Steiger J, Bopp A, Hoerstrup SP, Emmert MY. Translational cardiac stem cell therapy: advancing from first-generation to next-generation cell types. NPJ Regen Med. 2017;2(1):17.

30. Hastings CL, Roche ET, Ruiz-Hernandez E, Schenke-Layland K, Walsh CJ, Duffy GP. Drug and cell delivery for cardiac regeneration. Adv Drug Deliv Rev. 2015;84:85–106.

31. Frantz C, Stewart KM, Weaver VM. The extracellular matrix at a glance. J Cell Sci. 2010;123(24):4195.

32. Li Z, Masumoto H, Jo JI, Yamazaki K, Ikeda T, Tabata Y, Minatoya K. Sustained release of basic fibroblast growth factor using gelatin hydrogel improved left ventricular function through the alteration of collagen subtype in a rat chronic myocardial infarction model. Gen Thorac Cardiovasc Surg. 2018;66(11):641–7.

33. Kumagai M, Minakata K, Masumoto H, Yamamoto M, Yonezawa A, Ikeda T, Uehara K, Yamazaki K, Ikeda T, Matsubara K, et al. A therapeutic angiogenesis of sustained release of basic fibroblast growth factor using biodegradable gelatin hydrogel sheets in a canine chronic myocardial infarction model. Heart Vessels. 2018;33(10):1251-7.

34. Qu H, Xie B-D, Wu J, Lv B, Chuai J-B, Li J-Z, Cai J, Wu H, Jiang S-L, Leng X-P, et al. Improved left ventricular aneurysm repair with cell- and cytokine-seeded collagen patches. Stem Cells Int. 2018;2018:4717802.

35. O'Neill HS, O'Sullivan J, Porteous N, Ruiz-Hernandez E, Kelly HM, O'Brien FJ, Duffy GP. A collagen cardiac patch incorporating alginate microparticles permits the controlled release of hepatocyte growth factor and insulin-like growth factor-1 to enhance cardiac stem cell migration and proliferation. J Tissue Eng Regen Med. 2018;12(1):e384–94.

36. Pandey R, Velasquez S, Durrani S, Jiang M, Neiman M, Crocker JS, Benoit JB, Rubinstein J, Paul A, Ahmed RP. MicroRNA-1825 induces proliferation of adult cardiomyocytes and promotes cardiac regeneration post ischemic injury. Am J Transl Res. 2017;9(6):3120–37.

37. Wang LL, Liu Y, Chung JJ, Wang T, Gaffey AC, Lu M, Cavanaugh CA, Zhou S, Kanade R, Atluri P, et al. Local and sustained miRNA delivery from an injectable hydrogel promotes cardiomyocyte proliferation and functional regeneration after ischemic injury. Nat Biomed Eng. 2017;1:983–92.

38. Monaghan MG, Holeiter M, Brauchle E, Layland SL, Lu Y, Deb A, Pandit A, Nsair A, Schenke-Layland K. Exogenous miR-29B delivery through a hyaluronan-based injectable system yields functional maintenance of the infarcted myocardjum. Tissue Eng Part A. 2017;24(1–2):57–67.

39. Wang Y, Zhang J, Qin Z, Fan Z, Lu C, Chen B, Zhao J, Li X, Xiao F, Lin X, et al. Preparation of high bioactivity multilayered bone-marrow mesenchymal stem cell sheets for myocardial infarction using a 3D-dynamic system. Acta Biomater. 2018;72:182–95.

40. Paul A, Hasan A, Kindi HA, Gaharwar AK, Rao VTS, Nikkhah M, Shin SR, Krafft D, Dokmeci MR, Shum-Tim D, et al. Injectable graphene oxide/hydrogel-based angiogenic gene delivery system for vasculogenesis and cardiac repair. ACS Nano. 2014;8(8):8050–62.

41. Purcell BP, Barlow SC, Perreault PE, Freeburg L, Doviak H, Jacobs J, Hoenes A, Zellars KN, Khakoo AY, Lee T, et al. Delivery of a matrix metalloproteinase-responsive hydrogel releasing TIMP-3 after myocardial infarction: effects on left ventricular remodeling. Am J Physiol Heart Circ Physiol. 2018;315(4):H814–25.

42. Fan Z, Xu Z, Niu H, Gao N, Guan Y, Li C, Dang Y, Cui X, Liu XL, Duan Y, et al. An injectable oxygen release system to augment cell survival and promote cardiac repair following myocardial infarction. Sci Rep. 2018;8(1):1371.

43. Waters R, Alam P, Pacelli S, Chakravarti AR, Ahmed RPH, Paul A. Stem cell-inspired secretome-rich injectable hydrogel to repair injured cardiac tissue. Acta Biomater. 2018;69:95–106.

44. Atala A. Tissue engineering and regenerative medicine: concepts for clinical application. Rejuvenation Res. 2004;7(1):15–31.

45. Hasirci V, Yucel D. Polymers used in tissue engineering. In: Wnek GE, Bowlin GL, editors. Encyclopedia of biomaterials and biomedical engineering. New York: Informa Healthcare USA, Inc.; 2008. p. 2282–99.

46. Boland ED, Espy PG, Bowlin GL. Tissue engineering scaffolds. In: Wnek GE, Bowlin GL, editors. Encyclopedia of biomaterials and biomedical engineering. New York: Informa Healthcare USA, Inc.; 2008. p. 2828 37.
47. Jang J, Park H-J, Kim S-W, Kim H, Park JY, Na SJ, Kim HJ, Park MN, Choi SH, Park SH, et al. 3D printed complex tissue construct using stem cell-laden decellularized extracellular matrix bioinks for cardiac repair. Biomaterials. 2017;112:264–74.
48. French KM, Somasuntharam I, Davis ME. Self-assembling peptide-based delivery of therapeutics for myocardial infarction. Adv Drug Deliv Rev. 2016;96:40–53.
49. Norahan MH, Amroon M, Ghahremanzadeh R, Mahmoodi M, Baheiraei N. Electroactive graphene oxide-incorporated collagen assisting vascularization for cardiac tissue engineering. J Biomed Mater Res A. 2019;107(1):204–19.
50. Hosoyama K, Ahumada M, McTiernan CD, Davis DR, Variola F, Ruel M, Liang W, Suuronen EJ, Alarcon EI. Nanoengineered electroconductive collagen-based cardiac patch for infarcted myocardium repair. ACS Appl Mater Interfaces. 2018;10(51):44668-77.
51. Dvir T, Timko BP, Brigham MD, Naik SR, Karajanagi SS, Levy O, Jin II, Parker KK, Langer R, Kohane DS. Nanowired three-dimensional cardiac patches. Nat Nanotechnol. 2011;6(11):720–5.
52. Sun H, Zhou J, Huang Z, Qu L, Lin N, Liang C, Dai R, Tang L, Tian F. Carbon nanotube-incorporated collagen hydrogels improve cell alignment and the performance of cardiac constructs. Int J Nanomed. 2017;12:3109–20.
53. Sherrell PC, Cieślar-Pobuda A, Ejneby MS, Sammalisto L, Gelmi A, de Muinck E, Brask J, Łos MJ, Rafat M. Rational design of a conductive collagen heart patch. Macromol Biosci. 2017;17(7):1600446.
54. Kai D, Prabhakaran MP, Jin G, Ramakrishna S. Biocompatibility evaluation of electrically conductive nanofibrous scaffolds for cardiac tissue engineering. J Mater Chem B. 2013;1(17):2305–14.
55. He Y, Ye G, Song C, Li C, Xiong W, Yu L, Qiu X, Wang L. Mussel-inspired conductive nanofibrous membranes repair myocardial infarction by enhancing cardiac function and revascularization. Theranostics. 2018;8(18):5159–77.
56. Dong R, Zhao X, Guo B, Ma PX. Self-healing conductive injectable hydrogels with antibacterial activity as cell delivery carrier for cardiac cell therapy. ACS Appl Mater Interfaces. 2016;8(27):17138–50.
57. Bao R, Tan B, Liang S, Zhang N, Wang W, Liu W. A π-π conjugation-containing soft and conductive injectable polymer hydrogel highly efficiently rebuilds cardiac function after myocardial infarction. Biomaterials. 2017;122:63–71.
58. Mihic A, Cui Z, Wu J, Vlacic G, Miyagi Y, Li S-H, Lu S, Sung H-W, Weisel Richard D, Li R-K. A conductive polymer hydrogel supports cell electrical signaling and improves cardiac function after implantation into myocardial infarct. Circulation. 2015;132(8):772–84.
59. Saravanan S, Sareen N, Abu-El-Rub E, Ashour H, Sequiera GL, Ammar HI, Gopinath V, Shamaa AA, Sayed SSE, Moudgil M, et al. Graphene oxide-gold nanosheets containing chitosan scaffold improves ventricular contractility and function after implantation into infarcted heart. Sci Rep. 2018;8(1):15069.
60. Kapnisi M, Mansfield C, Marijon C, Guex AG, Perbellini F, Bardi I, Humphrey EJ, Puetzer JL, Mawad D, Koutsogeorgis DC, et al. Auxetic cardiac patches with tunable mechanical and conductive properties toward treating myocardial infarction. Adv Funct Mater. 2018;28(21):1800618.
61. Zhou J, Yang X, Liu W, Wang C, Shen Y, Zhang F, Zhu H, Sun H, Chen J, Lam J, et al. Injectable OPF/graphene oxide hydrogels provide mechanical support and enhance cell electrical signaling after implantation into myocardial infarct. Theranostics. 2018;8(12):3317–30.
62. Christman KL, Lee RJ. Biomaterials for the treatment of myocardial infarction. J Am Coll Cardiol. 2006;48(5):907–13.
63. Domenech M, Polo-Corrales L, Ramirez-Vick JE, Freytes DO. Tissue engineering strategies for myocardial regeneration: acellular versus cellular scaffolds? Tissue Eng Part B Rev. 2016;22(6):438–58.

64. Ye L, Zimmermann W-H, Garry Daniel J, Zhang J. Patching the heart. Circ Res. 2013;113(7):922–32.
65. Wang Q-l, Wang H-j, Li Z-h, Wang Y-l, Wu X-p, Tan Y-z. Mesenchymal stem cell-loaded cardiac patch promotes epicardial activation and repair of the infarcted myocardium. J Cell Mol Med. 2017;21(9):1751–66.
66. Suarez SL, Rane AA, Muñoz A, Wright AT, Zhang SX, Braden RL, Almutairi A, McCulloch AD, Christman KL. Intramyocardial injection of hydrogel with high interstitial spread does not impact action potential propagation. Acta Biomater. 2015;26:13–22.
67. Qiu Y, Hamilton SK, Temenoff J. Improving mechanical properties of injectable polymers and composites, chap. 4. In: Vernon B, editor. Injectable biomaterials. Woodhead Publishing; 2011. p. 61–91.
68. Lakshmanan R, Kumaraswamy P, Krishnan UM, Sethuraman S. Engineering a growth factor embedded nanofiber matrix niche to promote vascularization for functional cardiac regeneration. Biomaterials. 2016;97:176–95.
69. Sadtler K, Singh A, Wolf MT, Wang X, Pardoll DM, Elisseeff JH. Design, clinical translation and immunological response of biomaterials in regenerative medicine. Nat Rev Mater. 2016;1:16040.
70. Drury JL, Mooney DJ. Hydrogels for tissue engineering: scaffold design variables and applications. Biomaterials. 2003;24(24):4337–51.
71. Lee JH. Injectable hydrogels delivering therapeutic agents for disease treatment and tissue engineering. Biomater Res. 2018;22(1):27.
72. Lee KY, Mooney DJ. Hydrogels for tissue engineering. Chem Rev. 2001;101(7):1869–80.
73. Rodell CB, Kaminski AL, Burdick JA. Rational design of network properties in guest-host assembled and shear-thinning hyaluronic acid hydrogels. Biomacromolecules. 2013;14(11):4125–34.
74. Rodell CB, MacArthur JW Jr, Dorsey SM, Wade RJ, Wang LL, Woo YJ, Burdick JA. Shear-thinning supramolecular hydrogels with secondary autonomous covalent crosslinking to modulate viscoelastic properties in vivo. Adv Funct Mater. 2015;25(4):636–44.
75. Wong Po Foo CTS, Lee JS, Mulyasasmita W, Parisi-Amon A, Heilshorn SC. Two-component protein-engineered physical hydrogels for cell encapsulation. PNAS 2009;106(52):22067–72.
76. Lu HD, Charati MB, Kim IL, Burdick JA. Injectable shear-thinning hydrogels engineered with a self-assembling Dock-and-Lock mechanism. Biomaterials. 2012;33(7):2145–53.
77. Liu Y, Hsu S-h. Synthesis and biomedical applications of self-healing hydrogels. Front Chem. 2018;6(449).
78. Lindsey ML, Bolli R, Canty JM Jr, Du X-J, Frangogiannis NG, Frantz S, Gourdie RG, Holmes JW, Jones SP, Kloner RA, et al. Guidelines for experimental models of myocardial ischemia and infarction. Am J Physiol Heart Circ Physiol. 2018;314(4):H812–38.
79. Naderi H, Matin MM, Bahrami AR. Review paper: critical issues in tissue engineering: biomaterials, cell sources, angiogenesis, and drug delivery systems. J Biomater Appl. 2011;26(4):383–417.
80. Nair LS, Laurencin CT. Biodegradable polymers as biomaterials. Prog Polym Sci. 2007;32(8):762–98.
81. Wang T, Lew J, Premkumar J, Poh CL, Naing MW. Production of recombinant collagen: state of the art and challenges. Eng Biol. 2017;1(1):18–23.
82. Blackburn NJR, Sofrenovic T, Kuraitis D, Ahmadi A, McNeill B, Deng C, Rayner KJ, Zhong Z, Ruel M, Suuronen EJ. Timing underpins the benefits associated with injectable collagen biomaterial therapy for the treatment of myocardial infarction. Biomaterials. 2015;39:182–92.
83. Serpooshan V, Zhao M, Metzler SA, Wei K, Shah PB, Wang A, Mahmoudi M, Malkovskiy AV, Rajadas J, Butte MJ, et al. The effect of bioengineered acellular collagen patch on cardiac remodeling and ventricular function post myocardial infarction. Biomaterials. 2013;34(36):9048–55.
84. Dai W, Wold LE, Dow JS, Kloner RA. Thickening of the infarcted wall by collagen injection improves left ventricular function in rats: a novel approach to preserve cardiac function after myocardial infarction. J Am Coll Cardiol. 2005;46(4):714–9.

85. Joshi J, Brennan D, Beachley V, Kothapalli CR. Cardiomyogenic differentiation of human bone marrow-derived mesenchymal stem cell spheroids within electrospun collagen nanofiber mats. J Biomed Mater Res A. 2018;106(12):3303–12.

86. Valarmathi MT, Fuseler JW, Davis JM, Price RL. A novel human tissue-engineered 3-D functional vascularized cardiac muscle construct. Front Cell Dev Biol. 2017;5(2).

87. Ahmadi A, McNeill B, Vulesevic B, Kordos M, Mesana L, Thorn S, Renaud JM, Manthorp E, Kuraitis D, Toeg H, et al. The role of integrin α2 in cell and matrix therapy that improves perfusion, viability and function of infarcted myocardium. Biomaterials. 2014;35(17):4749–58.

88. Giordano C, Thorn Stephanie L, Renaud Jennifer M, Al-Atassi T, Boodhwani M, Klein R, Kuraitis D, Dwivedi G, Zhang P, DaSilva Jean N, et al. Preclinical evaluation of biopolymer-delivered circulating angiogenic cells in a swine model of hibernating myocardium. Circ Cardiovasc Imaging. 2013;6(6):982–91.

89. Xu G, Wang X, Deng C, Teng X, Suuronen EJ, Shen Z, Zhong Z. Injectable biodegradable hybrid hydrogels based on thiolated collagen and oligo(acryloyl carbonate)–poly(ethylene glycol)–oligo(acryloyl carbonate) copolymer for functional cardiac regeneration. Acta Biomater. 2015;15:55–64.

90. Xia Y, Zhu K, Lai H, Lang M, Xiao Y, Lian S, Guo C, Wang C. Enhanced infarct myocardium repair mediated by thermosensitive copolymer hydrogel-based stem cell transplantation. Exp Biol Med (Maywood). 2015;240(5):593–600.

91. Liu Y, Xu Y, Wang Z, Wen D, Zhang W, Schmull S, Li H, Chen Y, Xue S. Electrospun nanofibrous sheets of collagen/elastin/polycaprolactone improve cardiac repair after myocardial infarction. Am J Transl Res. 2016;8(4):1678–94.

92. Shafiq M, Zhang Y, Zhu D, Zhao Z, Kim D-H, Kim SH, Kong D. In situ cardiac regeneration by using neuropeptide substance P and IGF-1C peptide eluting heart patches. Regen Biomater. 2018;5(5):303–16.

93. Reis LA, Chiu LL, Wu J, Feric N, Laschinger C, Momen A, Li R-K, Radisic M. Hydrogels with integrin-binding angiopoietin-1-derived peptide, QHREDGS, for treatment of acute myocardial infarction. Circ Heart Fail. 2015;8(2):333–41.

94. Hosoyama K, Ahumada M, McTiernan CD, Bejjani J, Variola F, Ruel M, Xu B, Liang W, Suuronen EJ, Alarcon EI. Multi-functional thermo-crosslinkable collagen-metal nanoparticle composites for tissue regeneration: nanosilver vs. nanogold. RSC Adv. 2017;7(75):47704–8.

95. Chachques JC, Trainini JC, Lago N, Masoli OH, Barisani JL, Cortes-Morichetti M, Schussler O, Carpentier A. Myocardial assistance by grafting a new bioartificial upgraded myocardium (MAGNUM clinical trial): one year follow-up. Cell Transplant. 2007;16(9):927–34.

96. Latifi N, Asgari M, Vali H, Mongeau L. A tissue-mimetic nano-fibrillar hybrid injectable hydrogel for potential soft tissue engineering applications. Sci Rep. 2018;8(1):1047.

97. Su K, Wang C. Recent advances in the use of gelatin in biomedical research. Biotechnol Lett. 2015;37(11):2139–45.

98. Nakajima K, Fujita J, Matsui M, Tohyama S, Tamura N, Kanazawa H, Seki T, Kishino Y, Hirano A, Okada M, et al. Gelatin hydrogel enhances the engraftment of transplanted cardiomyocytes and angiogenesis to ameliorate cardiac function after myocardial infarction. PLoS One. 2015;10(7):e0133308.

99. Zhou J, Chen J, Sun H, Qiu X, Mou Y, Liu Z, Zhao Y, Li X, Han Y, Duan C, et al. Engineering the heart: evaluation of conductive nanomaterials for improving implant integration and cardiac function. Sci Rep. 2014;4:3733.

100. Shin SR, Jung SM, Zalabany M, Kim K, Zorlutuna P, Kim Sb, Nikkhah M, Khabiry M, Azize M, Kong J, et al. Carbon-nanotube-embedded hydrogel sheets for engineering cardiac constructs and bioactuators. ACS Nano. 2013;7(3):2369–80.

101. Noshadi I, Hong S, Sullivan KE, Shirzaei Sani E, Portillo-Lara R, Tamayol A, Shin SR, Gao AE, Stoppel WL, Black Iii LD, et al. In vitro and in vivo analysis of visible light crosslinkable gelatin methacryloyl (GelMA) hydrogels. Biomater Sci. 2017;5(10):2093–105.

102. Yang B, Yao F, Hao T, Fang W, Ye L, Zhang Y, Wang Y, Li J, Wang C. Development of electrically conductive double-network hydrogels via one-step facile strategy for cardiac tissue engineering. Adv Healthcare Mater. 2016;5:474–88.

103. Wang W, Tao H, Zhao Y, Sun X, Tang J, Selomulya C, Tang J, Chen T, Wang Y, Shu M, et al. Implantable and biodegradable macroporous iron oxide frameworks for efficient regeneration and repair of infracted heart. Theranostics. 2017;7(7):1966–75.
104. Chang M-Y, Huang T-T, Chen C-H, Cheng B, Hwang S-M, Hsieh PCH. Injection of human cord blood cells with hyaluronan improves postinfarction cardiac repair in pigs. Stem Cells Transl Med. 2016;5(1):56–66.
105. Vu TD, Pal SN, Ti L-K, Martinez EC, Rufaihah AJ, Ling LH, Lee C-N, Richards AM, Kofidis T. An autologous platelet-rich plasma hydrogel compound restores left ventricular structure, function and ameliorates adverse remodeling in a minimally invasive large animal myocardial restoration model: a translational approach: Vu and Pal "myocardial repair: PRP, hydrogel and supplements". Biomaterials. 2015;45:27–35.
106. Rodell CB, Lee ME, Wang H, Takebayashi S, Takayama T, Kawamura T, Arkles JS, Dusaj NN, Dorsey SM, Witschey WRT, et al. Injectable shear-thinning hydrogels for minimally invasive delivery to infarcted myocardium to limit left ventricular remodeling. Circ Cardiovasc Interv. 2016;9(10):e004058.
107. Wang LL, Chung JJ, Li EC, Uman S, Atluri P, Burdick JA. Injectable and protease-degradable hydrogel for siRNA sequestration and triggered delivery to the heart. J Control Release. 2018;285:152–61.
108. Chen CW, Wang LL, Zaman S, Gordon J, Arisi MF, Venkataraman CM, Chung JJ, Hung G, Gaffey AC, Spruce LA, et al. Sustained release of endothelial progenitor cell-derived extracellular vesicles from shear-thinning hydrogels improves angiogenesis and promotes function after myocardial infarction. Cardiovasc Res. 2018;114(7):1029–40.
109. Fan Z, Fu M, Xu Z, Zhang B, Li Z, Li H, Zhou X, Liu X, Duan Y, Lin P-H, et al. Sustained release of a peptide-based matrix metalloproteinase-2 inhibitor to attenuate adverse cardiac remodeling and improve cardiac function following myocardial infarction. Biomacromolecules. 2017;18(9):2820–9.
110. Wang W, Tan B, Chen J, Bao R, Zhang X, Liang S, Shang Y, Liang W, Cui Y, Fan G, et al. An injectable conductive hydrogel encapsulating plasmid DNA-eNOs and ADSCs for treating myocardial infarction. Biomaterials. 2018;160:69–81.
111. Dorsey SM, McGarvey JR, Wang H, Nikou A, Arama L, Koomalsingh KJ, Kondo N, Gorman JH, Pilla JJ, Gorman RC, et al. MRI evaluation of injectable hyaluronic acid-based hydrogel therapy to limit ventricular remodeling after myocardial infarction. Biomaterials. 2015;69:65–75.
112. Tang J, Vandergriff A, Wang Z, Hensley MT, Cores J, Allen TA, Dinh P-U, Zhang J, Caranasos TG, Cheng K. A regenerative cardiac patch formed by spray painting of biomaterials onto the heart. Tissue Eng Part C. 2017;23(3):146–55.
113. Vallée J-P, Hauwel M, Lepetit-Coiffé M, Bei W, Montet-Abou K, Meda P, Gardier S, Zammaretti P, Kraehenbuehl TP, Herrmann F, et al. Embryonic stem cell-based cardiopatches improve cardiac function in infarcted rats. Stem Cells Transl Med. 2012;1(3):248–60.
114. Xiong Q, Hill KL, Li Q, Suntharalingam P, Mansoor A, Wang X, Jameel MN, Zhang P, Swingen C, Kaufman DS, et al. A fibrin patch-based enhanced delivery of human embryonic stem cell-derived vascular cell transplantation in a porcine model of postinfarction left ventricular remodeling. Stem Cells. 2011;29(2):367–75.
115. Bellamy V, Vanneaux V, Bel A, Nemetalla H, Emmanuelle Boitard S, Farouz Y, Joanne P, Perier M-C, Robidel E, Mandet C, et al. Long-term functional benefits of human embryonic stem cell-derived cardiac progenitors embedded into a fibrin scaffold. J Heart Lung Transplant. 2015;34(9):1198–207.
116. Blondiaux E, Pidial L, Autret G, Rahmi G, Balvay D, Audureau E, Wilhelm C, Guerin CL, Bruneval P, Silvestre J-S, et al. Bone marrow-derived mesenchymal stem cell-loaded fibrin patches act as a reservoir of paracrine factors in chronic myocardial infarction. J Tissue Eng Regen Med. 2017;11(12):3417–27.
117. Wendel JS, Ye L, Tao R, Zhang J, Zhang J, Kamp TJ, Tranquillo RT. Functional effects of a tissue-engineered cardiac patch from human induced pluripotent stem cell-derived cardiomyocytes in a rat infarct model. Stem Cells Transl Med. 2015;4(11):1324–32.

118. Mattapally S, Zhu W, Fast VG, Gao L, Worley C, Kannappan R, Borovjagin AV, Zhang J. Spheroids of cardiomyocytes derived from human induced pluripotent stem cells improve recovery from myocardial injury in mice. Am J Physiol Heart Circ Physiol. 2018;315(2):H327–39.

119. Roura S, Gálvez-Montón C, Bayes-Genis A. Fibrin, the preferred scaffold for cell transplantation after myocardial infarction? An old molecule with a new life. J Tissue Eng Regen Med. 2017;11(8):2304–13.

120. Higuchi A, Ku N-J, Tseng Y-C, Pan C-H, Li H-F, Kumar SS, Ling Q-D, Chang Y, Alarfaj AA, Munusamy MA, et al. Stem cell therapies for myocardial infarction in clinical trials: bioengineering and biomaterial aspects. Lab Invest. 2017;97:1167.

121. Menasché P, Vanneaux V, Hagège A, Bel A, Cholley B, Cacciapuoti I, Parouchev A, Benhamouda N, Tachdjian G, Tosca L, et al. Human embryonic stem cell-derived cardiac progenitors for severe heart failure treatment: first clinical case report. Eur Heart J. 2015;36(30):2011–7.

122. Gao L, Gregorich Zachery R, Zhu W, Mattapally S, Oduk Y, Lou X, Kannappan R, Borovjagin Anton V, Walcott Gregory P, Pollard Andrew E, et al. Large cardiac muscle patches engineered from human induced-pluripotent stem cell-derived cardiac cells improve recovery from myocardial infarction in swine. Circulation. 2018;137(16):1712–30.

123. Riemenschneider SB, Mattia DJ, Wendel JS, Schaefer JA, Ye L, Guzman PA, Tranquillo RT. Inosculation and perfusion of pre-vascularized tissue patches containing aligned human microvessels after myocardial infarction. Biomaterials. 2016;97:51–61.

124. Morin KT, Dries-Devlin JL, Tranquillo RT. Engineered microvessels with strong alignment and high lumen density via cell-induced fibrin gel compaction and interstitial flow. Tissue Eng Part A. 2013;20(3–4):553–65.

125. Su T, Huang K, Daniele MA, Hensley MT, Young AT, Tang J, Allen TA, Vandergriff AC, Erb PD, Ligler FS, et al. Cardiac stem cell patch integrated with microengineered blood vessels promotes cardiomyocyte proliferation and neovascularization after acute myocardial infarction. ACS Appl Mater Interfaces. 2018;10(39):33088–96.

126. Wang Z, Lee SJ, Cheng H-J, Yoo JJ, Atala A. 3D bioprinted functional and contractile cardiac tissue constructs. Acta Biomater. 2018;70:48–56.

127. Rufaihah AJ, Johari NA, Vaibavi SR, Plotkin M, Di Thien DT, Kofidis T, Seliktar D. Dual delivery of VEGF and ANG-1 in ischemic hearts using an injectable hydrogel. Acta Biomater. 2017;48:58–67.

128. Bearzi C, Gargioli C, Baci D, Fortunato O, Shapira-Schweitzer K, Kossover O, Latronico MVG, Seliktar D, Condorelli G, Rizzi R. PlGF-MMP9-engineered iPS cells supported on a PEG-fibrinogen hydrogel scaffold possess an enhanced capacity to repair damaged myocardium. Cell Death Dis. 2014;5(2):e1053.

129. Spotnitz WD. Fibrin sealant: the only approved hemostat, sealant, and adhesive—a laboratory and clinical perspective. ISRN Surg. 2014;2014:203943.

130. Ott HC, Matthiesen TS, Goh S-K, Black LD, Kren SM, Netoff TI, Taylor DA. Perfusion-decellularized matrix: using nature's platform to engineer a bioartificial heart. Nat Med. 2008;14:213.

131. Wassenaar JW, Gaetani R, Garcia JJ, Braden RL, Luo CG, Huang D, DeMaria AN, Omens JH, Christman KL. Evidence for mechanisms underlying the functional benefits of a myocardial matrix hydrogel for post-MI treatment. J Am Coll Cardiol. 2016;67(9):1074–86.

132. Sarig U, Sarig H, de-Berardinis E, Chaw S-Y, Nguyen EBV, Ramanujam VS, Thang VD, Al-Haddawi M, Liao S, Seliktar D, et al. Natural myocardial ECM patch drives cardiac progenitor based restoration even after scarring. Acta Biomater. 2016;44:209–20.

133. Shah M, Kc P, Copeland KM, Liao J, Zhang G. A thin layer of decellularized porcine myocardium for cell delivery. Sci Rep. 2018;8(1):16206.

134. Wang Q, Yang H, Bai A, Jiang W, Li X, Wang X, Mao Y, Lu C, Qian R, Guo F, et al. Functional engineered human cardiac patches prepared from nature's platform improve heart function after acute myocardial infarction. Biomaterials. 2016;105:52–65.

135. Kajbafzadeh AM, Tafti SHA, Khorramirouz R, Sabetkish S, Kameli SM, Orangian S, Rabbani S, Oveisi N, Golmohammadi M, Kashani Z. Evaluating the role of autologous mesenchymal stem cell seeded on decellularized pericardium in the treatment of myocardial infarction: an animal study. Cell Tissue Bank. 2017;18(4):527–38.
136. Kameli SM, Khorramirouz R, Eftekharzadeh S, Fendereski K, Daryabari SS, Tavangar SM, Kajbafzadeh A-M. Application of tissue-engineered pericardial patch in rat models of myocardial infarction. J Biomed Mater Res A. 2018;106(10):2670–8.
137. Wan L, Chen Y, Wang Z, Wang W, Schmull S, Dong J, Xue S, Imboden H, Li J. Human heart valve-derived scaffold improves cardiac repair in a murine model of myocardial infarction. Sci Rep. 2017;7:39988.
138. Li N, Huang R, Zhang X, Xin Y, Li J, Huang Y, Cui W, Stoltz JF, Zhou Y, Kong Q. Stem cells cardiac patch from decellularized umbilical artery improved heart function after myocardium infarction. Biomed Mater Eng. 2017;28(s1):S87–94.
139. Toeg HD, Tiwari-Pandey R, Seymour R, Ahmadi A, Crowe S, Vulesevic B, Suuronen EJ, Ruel M. Injectable small intestine submucosal extracellular matrix in an acute myocardial infarction model. Ann Thorac Surg. 2013;96(5):1686–94.
140. Sonnenberg SB, Rane AA, Liu CJ, Rao N, Agmon G, Suarez S, Wang R, Munoz A, Bajaj V, Zhang S, et al. Delivery of an engineered HGF fragment in an extracellular matrix-derived hydrogel prevents negative LV remodeling post-myocardial infarction. Biomaterials. 2015;45:56–63.
141. Francis MP, Breathwaite E, Bulysheva AA, Varghese F, Rodriguez RU, Dutta S, Semenov I, Ogle R, Huber A, Tichy A-M, et al. Human placenta hydrogel reduces scarring in a rat model of cardiac ischemia and enhances cardiomyocyte and stem cell cultures. Acta Biomater. 2017;52:92–104.
142. Mewhort HEM, Turnbull JD, Satriano A, Chow K, Flewitt JA, Andrei A-C, Guzzardi DG, Svystonyuk DA, White JA, Fedak PWM. Epicardial infarct repair with bioinductive extracellular matrix promotes vasculogenesis and myocardial recovery. J Heart Lung Transplant. 2016;35(5):661–70.
143. Mewhort HEM, Svystonyuk DA, Turnbull JD, Teng G, Belke DD, Guzzardi DG, Park DS, Kang S, Hollenberg MD, Fedak PWM. Bioactive extracellular matrix scaffold promotes adaptive cardiac remodeling and repair. JACC Basic Transl Sci. 2017;2(4):450–64.
144. Wang L, Meier EM, Tian S, Lei I, Liu L, Xian S, Lam MT, Wang Z. Transplantation of Isl1 + cardiac progenitor cells in small intestinal submucosa improves infarcted heart function. Stem Cell Res Ther. 2017;8(1):230.
145. Soucy KG, Smith EF, Monreal G, Rokosh G, Keller BB, Yuan F, Matheny RG, Fallon AM, Lewis BC, Sherwood LC, et al. Feasibility study of particulate extracellular matrix (P-ECM) and left ventricular assist device (HVAD) therapy in chronic ischemic heart failure bovine model. ASAIO J. 2015;61(2):161–9.
146. Calgary Uo. Epicardial infarct repair using CorMatrix®-ECM: clinical feasibility study (EIR). National Library of Medicine. https://www.clinicaltrials.gov/ct2/show/NCT02887768?term= NCT02887768&rank=1 (2016). Accessed 07 Jan 2019.
147. Singelyn JM, DeQuach JA, Seif-Naraghi SB, Littlefield RB, Schup-Magoffin PJ, Christman KL. Naturally derived myocardial matrix as an injectable scaffold for cardiac tissue engineering. Biomaterials. 2009;30(29):5409–16.
148. Singelyn JM, Sundaramurthy P, Johnson TD, Schup-Magoffin PJ, Hu DP, Faulk DM, Wang J, Mayle KM, Bartels K, Salvatore M, et al. Catheter-deliverable hydrogel derived from decellularized ventricular extracellular matrix increases endogenous cardiomyocytes and preserves cardiac function post-myocardial infarction. J Am Coll Cardiol. 2012;59(8):751–63.
149. Seif-Naraghi SB, Singelyn JM, Salvatore MA, Osborn KG, Wang JJ, Sampat U, Kwan OL, Strachan GM, Wong J, Schup-Magoffin PJ, et al. Safety and efficacy of an injectable extracellular matrix hydrogel for treating myocardial infarction. Sci Transl Med. 2013;5(173):173ra25.
150. Ventrix I. A phase I, open-label study of the effects of percutaneous administration of an extracellular matrix hydrogel, VentriGel, following myocardial infarction. National Library of Medicine. https://clinicaltrials.gov/ct2/show/NCT02305602?term=NCT02305602 (2015). Accessed 15 Feb 2019.

151. Zhu Y, Matsumura Y, Wagner WR. Ventricular wall biomaterial injection therapy after myocardial infarction: advances in material design, mechanistic insight and early clinical experiences. Biomaterials. 2017;129:37–53.
152. Panda NC, Zuckerman ST, Mesubi OO, Rosenbaum DS, Penn MS, Donahue JK, Alsberg E, Laurita KR. Improved conduction and increased cell retention in healed MI using mesenchymal stem cells suspended in alginate hydrogel. J Interv Card Electrophysiol. 2014;41(2):117–27.
153. Daskalopoulos EP, Vilaeti AD, Barka E, Mantzouratou P, Kouroupis D, Kontonika M, Tourmousoglou C, Papalois A, Pantos C, Blankesteijn WM, et al. Attenuation of post-infarction remodeling in rats by sustained myocardial growth hormone administration. Growth Factors 2015;33(4):250–8.
154. Kontonika M, Barka E, Roumpi M, La Rocca V, Lekkas P, Daskalopoulos EP, Vilaeti AD, Baltogiannis GG, Vlahos AP, Agathopoulos S, et al. Prolonged intra-myocardial growth hormone administration ameliorates post-infarction electrophysiologic remodeling in rats. Growth Factors 2017;35(1):1–11.
155. Fang R, Qiao S, Liu Y, Meng Q, Chen X, Song B, Hou X, Tian W. Sustained co-delivery of BIO and IGF-1 by a novel hybrid hydrogel system to stimulate endogenous cardiac repair in myocardial infarcted rat hearts. Int J Nanomed. 2015;10:4691–703.
156. Rodness J, Mihic A, Miyagi Y, Wu J, Weisel RD, Li R-K. VEGF-loaded microsphere patch for local protein delivery to the ischemic heart. Acta Biomater. 2016;45:169–81.
157. Singh RD, Hillestad ML, Livia C, Li M, Alekseev AE, Witt TA, Stalboerger PG, Yamada S, Terzic A, Behfar A. M3RNA drives targeted gene delivery in acute myocardial infarction. Tissue Eng Part A. 2018;25(1-2):145-58.
158. Leor J, Tuvia S, Guetta V, Manczur F, Castel D, Willenz U, Petneházy Ö, Landa N, Feinberg MS, Konen E, et al. Intracoronary injection of in situ forming alginate hydrogel reverses left ventricular remodeling after myocardial infarction in swine. J Am Coll Cardiol. 2009;54(11):1014–23.
159. Frey N, Linke A, Süselbeck T, Müller-Ehmsen J, Vermeersch P, Schoors D, Rosenberg M, Bea F, Tuvia S, Leor J. Intracoronary delivery of injectable bioabsorbable scaffold (IK-5001) to treat left ventricular remodeling after ST-elevation myocardial infarction. Circ Cardiovasc Interv. 2014;7(6):806–12.
160. Rao SV, Zeymer U, Douglas PS, Al-Khalidi H, Liu J, Gibson CM, Harrison RW, Joseph DS, Heyrman R, Krucoff MW. A randomized, double-blind, placebo-controlled trial to evaluate the safety and effectiveness of intracoronary application of a novel bioabsorbable cardiac matrix for the prevention of ventricular remodeling after large ST-segment elevation myocardial infarction: rationale and design of the PRESERVATION I trial. Am Heart J. 2015;170(5):929–37.
161. McCune C, McKavanagh P, Menown IBA. A review of the key clinical trials of 2015: results and implications. Cardiol Ther. 2016;5(2):109–32.
162. Sabbah HN, Wang M, Gupta RC, Rastogi S, Ilsar I, Sabbah MS, Kohli S, Helgerson S, Lee RJ. Augmentation of left ventricular wall thickness with alginate hydrogel implants improves left ventricular function and prevents progressive remodeling in dogs with chronic heart failure. JACC Heart Fail. 2013;1(3):252–8.
163. Lee LC, Wall ST, Klepach D, Ge L, Zhang Z, Lee RJ, Hinson A, Gorman JH, Gorman RC, Guccione JM. Algisyl-LVR™ with coronary artery bypass grafting reduces left ventricular wall stress and improves function in the failing human heart. Int J Cardiol. 2013;168(3):2022–8.
164. Mann DL, Lee RJ, Coats AJS, Neagoe G, Dragomir D, Pusineri E, Piredda M, Bettari L, Kirwan B-A, Dowling R, et al. One-year follow-up results from AUGMENT-HF: a multicentre randomized controlled clinical trial of the efficacy of left ventricular augmentation with algisyl in the treatment of heart failure. Eur J Heart Fail. 2016;18(3):314–25.
165. Henning RJ, Khan A, Jimenez E. Chitosan hydrogels significantly limit left ventricular infarction and remodeling and preserve myocardial contractility. J Surg Res. 2016;201(2):490–7.
166. Xu B, Li Y, Deng B, Liu X, Wang L, Zhu Q-L. Chitosan hydrogel improves mesenchymal stem cell transplant survival and cardiac function following myocardial infarction in rats. Exp Ther Med. 2017;13(2):588–94.

167. Hardy B, Battler A, Weiss C, Kudasi O, Raiter A. Therapeutic angiogenesis of mouse hind limb ischemia by novel peptide activating GRP78 receptor on endothelial cells. Biochem Pharmacol. 2008;75(4):891–9.

168. Shu Y, Hao T, Yao F, Qian Y, Wang Y, Yang B, Li J, Wang C. RoY peptide-modified chitosan-based hydrogel to improve angiogenesis and cardiac repair under hypoxia. ACS Appl Mater Interfaces. 2015;7(12):6505–17.

169. Chen J, Zhan Y, Wang Y, Han D, Tao B, Luo Z, Ma S, Wang Q, Li X, Fan L, et al. Chitosan/silk fibroin modified nanofibrous patches with mesenchymal stem cells prevent heart remodeling post-myocardial infarction in rats. Acta Biomater. 2018;80:154–68.

170. Wang X, Wang L, Wu Q, Bao F, Yang H, Qiu X, Chang J. Chitosan/calcium silicate cardiac patch stimulates cardiomyocyte activity and myocardial performance after infarction by synergistic effect of bioactive ions and aligned nanostructure. ACS Appl Mater Interfaces. 2018;11(1):1449-68.

171. Cui Z, Ni NC, Wu J, Du GQ, He S, Yau TM, Weisel RD, Sung HW, Li RK. Polypyrrole-chitosan conductive biomaterial synchronizes cardiomyocyte contraction and improves myocardial electrical impulse propagation. Theranostics. 2018;8(10):2752–64.

172. Koutsopoulos S. Self-assembling peptide nanofiber hydrogels in tissue engineering and regenerative medicine: progress, design guidelines, and applications. J Biomed Mater Res A. 2016;104(4):1002–16.

173. Cui H, Webber MJ, Stupp SI. Self-assembly of peptide amphiphiles: from molecules to nanostructures to biomaterials. Biopolymers. 2010;94(1):1–18.

174. Li H, Gao J, Shang Y, Hua Y, Ye M, Yang Z, Ou C, Chen M. Folic acid derived hydrogel enhances the survival and promotes therapeutic efficacy of iPS cells for acute myocardial infarction. ACS Appl Mater Interfaces. 2018;10(29):24459–68

175. Ichihara Y, Kaneko M, Yamahara K, Koulouroudias M, Sato N, Uppal R, Yamazaki K, Saito S, Suzuki K. Self-assembling peptide hydrogel enables instant epicardial coating of the heart with mesenchymal stromal cells for the treatment of heart failure. Biomaterials. 2018;154:12–23.

176. Boopathy AV, Martinez MD, Smith AW, Brown ME, García AJ, Davis ME. Intramyocardial delivery of notch ligand-containing hydrogels improves cardiac function and angiogenesis following infarction. Tissue Eng Part A. 2015;21(17–18):2315–22.

177. Martínez-Ramos C, Rodríguez-Pérez E, Garnes MP, Chachques JC, Moratal D, Vallés-Lluch A, Monleón Pradas M. Design and assembly procedures for large-sized biohybrid scaffolds as patches for myocardial infarct. Tissue Eng Part C Methods. 2014;20(10):817–27.

178. Lamboni L, Gauthier M, Yang G, Wang Q. Silk sericin: a versatile material for tissue engineering and drug delivery. Biotechnol Adv. 2015;33(8):1855–67.

179. Song Y, Zhang C, Zhang J, Sun N, Huang K, Li H, Wang Z, Huang K, Wang L. An injectable silk sericin hydrogel promotes cardiac functional recovery after ischemic myocardial infarction. Acta Biomater. 2016;41:210–23.

180. Malki M, Fleischer S, Shapira A, Dvir T. Gold nanorod-based engineered cardiac patch for suture-free engraftment by near IR. Nano Lett. 2018;18(7):4069–73.

181. Rabbani S, Soleimani M, Imani M, Sahebjam M, Ghiaseddin A, Nassiri SM, Majd Ardakani J, Tajik Rostami M, Jalali A, Mousanassab B, et al. Regenerating heart using a novel compound and human Wharton jelly mesenchymal stem cells. Arch Med Res. 2017;48(3):228–37.

182. Chow A, Stuckey DJ, Kidher E, Rocco M, Jabbour RJ, Mansfield CA, Darzi A, Harding SE, Stevens MM, Athanasiou T. Human induced pluripotent stem cell-derived cardiomyocyte encapsulating bioactive hydrogels improve rat heart function post myocardial infarction. Stem Cell Rep. 2017;9(5):1415–22.

183. Melhem M, Jensen T, Reinkensmeyer L, Knapp L, Flewellyn J, Schook L. A hydrogel construct and fibrin-based glue approach to deliver therapeutics in a murine myocardial infarction model. J Vis Exp. 2015;100:e52562.

184. Rabbani S, Soleimani M, Sahebjam M, Imani M, Haeri A, Ghiaseddin A, Nassiri SM, Majd Ardakani J, Tajik Rostami M, Jalali A, et al. Simultaneous delivery of Wharton's jelly mesenchymal stem cells and insulin-like growth factor-1 in acute myocardial infarction. Iran J Pharm Res. 2018;17(2):426–41.

185. Steele AN, Cai L, Truong VN, Edwards BB, Goldstone AB, Eskandari A, Mitchell AC, Marquardt LM, Foster AA, Cochran JR, et al. A novel protein-engineered hepatocyte growth factor analog released via a shear-thinning injectable hydrogel enhances post-infarction ventricular function. Biotechnol Bioeng. 2017;114(10):2379–89.

186. Ciuffreda MC, Malpasso G, Chokoza C, Bezuidenhout D, Goetsch KP, Mura M, Pisano F, Davies NH, Gnecchi M. Synthetic extracellular matrix mimic hydrogel improves efficacy of mesenchymal stromal cell therapy for ischemic cardiomyopathy. Acta Biomater. 2018;70:71–83.

187. Woodruff MA, Hutmacher DW. The return of a forgotten polymer—polycaprolactone in the 21st century. Prog Polym Sci. 2010;35(10):1217–56.

188. Soler-Botija C, Bagó JR, Llucià-Valldeperas A, Vallés-Lluch A, Castells-Sala C, Martínez-Ramos C, Fernández-Muiños T, Chachques JC, Pradas MM, Semino CE, et al. Engineered 3D bioimplants using elastomeric scaffold, self-assembling peptide hydrogel, and adipose tissue-derived progenitor cells for cardiac regeneration. Am J Transl Res. 2014;6(3):291–301.

189. Chung T-W, Lo H-Y, Chou T-H, Chen J-H, Wang S-S. Promoting cardiomyogenesis of hBMSC with a forming self-assembly hBMSC microtissues/HA-GRGD/SF-PCL cardiac patch is mediated by the synergistic functions of HA-GRGD. Macromol Biosci. 2017;17(3):1600173.

190. Zhu H, Jiang X, Li X, Hu M, Wan W, Wen Y, He Y, Zheng X. Intramyocardial delivery of VEGF165 via a novel biodegradable hydrogel induces angiogenesis and improves cardiac function after rat myocardial infarction. Heart Vessels. 2016;31(6):963–75.

191. Zhu H, Li X, Yuan M, Wan W, Hu M, Wang X, Jiang X. Intramyocardial delivery of bFGF with a biodegradable and thermosensitive hydrogel improves angiogenesis and cardio-protection in infarcted myocardium. Exp Ther Med. 2017;14(4):3609–15.

192. Zeng X, Zou L, Levine RA, Guerrero JL, Handschumacher MD, Sullivan SM, Braithwaite GJC, Stone JR, Solis J, Muratoglu OK, et al. Efficacy of polymer injection for ischemic mitral regurgitation: persistent reduction of mitral regurgitation and attenuation of left ventricular remodeling. JACC Cardiovasc Interv. 2015;8(2):355–63.

193. Tang J, Wang J, Huang K, Ye Y, Su T, Qiao L, Hensley MT, Caranasos TG, Zhang J, Gu Z, et al. Cardiac cell-integrated microneedle patch for treating myocardial infarction. Sci Adv. 2018;4(11):eaat9365.

194. Jamaiyar A, Wan W, Ohanyan V, Enrick M, Janota D, Cumpston D, Song H, Stevanov K, Kolz CL, Hakobyan T, et al. Alignment of inducible vascular progenitor cells on a microbundle scaffold improves cardiac repair following myocardial infarction. Basic Res Cardiol. 2017;112(4):41.

195. Spadaccio C, Nappi F, De Marco F, Sedati P, Taffon C, Nenna A, Crescenzi A, Chello M, Trombetta M, Gambardella I, et al. Implantation of a poly-L-lactide GCSF-functionalized scaffold in a model of chronic myocardial infarction. J Cardiovasc Transl Res. 2017;10(1):47–65.

196. Zhu Y, Wood NA, Fok K, Yoshizumi T, Park DW, Jiang H, Schwartzman DS, Zenati MA, Uchibori T, Wagner WR, et al. Design of a coupled thermoresponsive hydrogel and robotic system for postinfarct biomaterial injection therapy. Ann Thorac Surg. 2016;102(3):780–6.

197. Zhu Y, Matsumura Y, Velayutham M, Foley LM, Hitchens TK, Wagner WR. Reactive oxygen species scavenging with a biodegradable, thermally responsive hydrogel compatible with soft tissue injection. Biomaterials. 2018;177:98–112.

198. Yoshizumi T, Zhu Y, Jiang H, D'Amore A, Sakaguchi H, Tchao J, Tobita K, Wagner WR. Timing effect of intramyocardial hydrogel injection for positively impacting left ventricular remodeling after myocardial infarction. Biomaterials. 2016;83:182–93.

199. Shiekh PA, Singh A, Kumar A. Oxygen-releasing antioxidant cryogel scaffolds with sustained oxygen delivery for tissue engineering applications. ACS Appl Mater Interfaces. 2018;10(22):18458–69.

Chapter 10
Regulatory Normative of Nanomaterials for Their Use in Biomedicine

Caitlin Lazurko, Manuel Ahumada, Emilio I. Alarcon and Erik Jacques

Abstract With nanomedicines increasing in market value and disruptive potential, a rapidly moving field such as this will require engaging in the difficult task of responsible management and the development of appropriate guidelines, which falls into the jurisdiction of governmental agencies. While each is influenced by the countries politics and demands of the people, there are shared goals of improving market success, risk assessment, and safety optimization. In this chapter, we describe the regulatory landscape with regards to nanomedicines in various countries. We first start with the world's nanotechnological leaders in North America, the European Union, and Asian and then discuss the notable strides taken by emerging countries where nanomedicines have caught the public eye.

10.1 Nanomedicine's Market and Disruptive Potential

Engineering materials at the nanoscale *a.k.a.*, nanoengineering has demonstrated to be able to modify the macroscopic properties of a large number of biomimetic 3D matrices, tissue scaffolds, and even synthetic polymers. With a US $\approx$ \$336 billion worth global market value, medical devices, including therapeutic materials for tissue and organ repair, presents an attractive niche where there is a lot of potential and room for producing intellectual properties (IP) with added value. Further, with the increasing life expectancies in developed countries; the health sciences are facing challenges that our society has never met before; meaning the need for therapeutics

C. Lazurko · E. I. Alarcon · E. Jacques (✉)
Division of Cardiac Surgery, University of Ottawa Heart Institute,
40 Ruskin Street, Ottawa, ON K1Y 4W7, Canada
e-mail: ejacq048@uottawa.ca

C. Lazurko · E. I. Alarcon
Department of Biochemistry, Microbiology, and Immunology, Faculty of Medicine,
University of Ottawa, Ottawa, ON K1H 8M5, Canada

M. Ahumada
Facultad de Ciencias, Centro de Nanotecnología Aplicada, Universidad Mayor,
Huechuraba, RM, Chile

© Springer Nature Switzerland AG 2019
E. I. Alarcon and M. Ahumada (eds.), *Nanoengineering Materials for Biomedical Uses*,
https://doi.org/10.1007/978-3-030-31261-9_10

and devices for chronic medical conditions linked to aging. Also, the demand for novel materials and technologies that repair and regenerate failing tissues and organs can potentially improve the quality of life of the elderly.

When looking at the relatively low ratio between the blockbuster number of scientific publications and the actual number of materials that have hit the market, there is evidence of a gap and voids in the regulatory normative for nanomaterials and their use in biomedicine. In this chapter, we will present and discuss the regulatory aspects of nanomaterials for biomedical devices and therapeutics. This is an area which is rapidly evolving, and the reader is encouraged to constantly consult on actualizations from the agencies cited in the following sections. The selection of the regions revised in this chapter was performed on the basis of those zones where literature and normative were readily available.

10.2   Regulatory Normative in North America

Nanotechnology holds great promise, but there are also concerns over the safety and regulation of these technologies. Nanomaterials interact differently than their bulk material, which is often an advantage; it makes it difficult to understand and predict the potential impacts and toxicity of these new materials and technologies. Thus, regulation is essential, but it is also a difficult task as a result of the diversity of the field. For example, terms including nanotechnology, nanomaterial, etc., are used for a variety of different applications [1–3]. Moreover, these materials cannot be regulated using traditional regulatory frameworks, especially since there is often a lack of reliable, systematic, and generalizable data regarding the safety of these materials [1–3]. Due to this lack of information, policymakers are often relying on perceived safety and risks. This has been done in the past with regulations surrounding stem cell research and genetically modified organisms, amongst others, where scientists act as experts in the field to provide information on the potential risks associated with the technology [2, 3]. Therefore, multiple agencies and experts need to work together to provide important information for the regulation of nanomaterials in North America and worldwide [2].

In the United States, the U.S. Food and Drug Administration (FDA) regulates nanomaterials that fall under existing statutory authority [4]. In 2006, the FDA developed the FDA Nanotechnology Task Force to create regulatory normatives that allow the safe and effective innovation of nanotechnologies in FDA-regulated products. They aim to address gaps in the knowledge and regulation surrounding nanotechnologies to decrease risks and adverse events associated with these technologies [5, 6]. The Task Force does not have a specific definition for nanotechnologies or nanomaterials, as they are attempting to take a "broadly inclusive approach" to nanotechnologies [6]. The Task Force states that as more information about the interactions of nanotechnologies becomes available, they may develop a formal definition [6]. The FDA says that it supports the production and innovation of nanotechnologies and has developed the FDA Nanotechnology Regulatory Science Research Plan to "provide coordinated leadership on regulatory science activities and issues related to FDA-regulated products that either contain nanomaterial or otherwise involve the application of nanotechnology" [4, 7]. They use a science-based approach for assessing the safety of the products but are limited by statutory authority. For example, where the FDA has the authority to conduct a premarket assessment of products that contain nanomaterials they will include the nanomaterials in their evaluation, but if the product does not require a premarket review, the FDA suggests a consultation [2, 4]. These products include cosmetics, supplements, and some food products [4]. Due to the uncertainty surrounding the safety of nanomaterials, mostly as a result of unique properties resulting from the nanoscale and high surface area to volume ratio, this has the potential to cause harm to the consumers due to the lack of federal regulation. While the FDA will complete post-market monitoring, much of the regulation is being placed on the manufacturers who must ensure "products meet all applicable legal requirements, including safety standards" [4]. This demonstrates that traditional regulatory frameworks cannot comprehensively regulate nanotechnologies [2]. The FDA has also stated that they collaborate with foreign counterparts to share information and knowledge surrounding the regulation of nanotechnologies [4]. This is important to ensure global health and help regulate the diverse field of nanotechnologies.

While there are limitations to the authority of the FDA in regulating nanotechnologies, they are developing tools to improve research surrounding nanotechnologies. For example, the FDA aims to strengthen the physio-chemical characterization of nanotechnologies through methods to analyze the safety, physical, and chemical properties. They aim to develop improved non-clinical models to assess the technologies in vitro and in vivo. Last, they aim to evaluate, improve, and standardize risk characterization, evaluation, and communication [7].

In addition to the FDA, the US Environmental Protection Agency (EPA) is also researching nanomaterials, and developing methods to better regulate nanotechnologies [8]. They define nanomaterials as "chemical substances that have structures with dimensions at the nanoscale" [9]. To ensure the safety of nanomaterials, the EPA is regulating nanomaterials under the Toxic Substances Control Act (TSCA) which includes an information gather rule, stating that companies who produce nanomaterials must inform the EPA of the chemical composition, production volume, and manufacturing methods of the materials, amongst others. Moreover, companies must notify the EPA before manufacturing new materials so the EPA can ensure their safety [9].

In Canada, Health Canada is responsible for the regulation of products, including nanomaterials [10]. Health Canada recognizes its role in providing a regulatory framework to protect the health of Canadians, minimize potential risks, and realize the benefits of nanotechnology [10]. Similar to the United States and the FDA, Health Canada is using existing legislation to regulate nanotechnologies; however, they recognize the possibility that new regulatory approaches may be necessary for adequate regulation of these technologies.

Health Canada helps protect and promote health by using existing legislative and regulatory frameworks to mitigate the potential health risks of nanomaterials and to help realize their health benefits. However, it is recognized that new approaches may be necessary for the future to keep pace with advances in this area as there is inadequate information on risks associated with nanomaterials at this time. [10]

Health Canada states that there are "*no regulations specific to nanotechnology-based health and food products*" [10], and that they rely on existing regulations which require a product safety assessment before the sale in Canada [10]. Once again, this demonstrates a gap in the regulation of nanotechnologies [2]. Similar to the United States, Canada uses a broad definition when defining nanomaterials and nanotechnologies to encompass the diversity of the field. This allows for greater regulation of a variety of products to ensure the safety of Canadians [10]. Similar to the United States once again, Canada does not have a specific definition of nanomaterials or nanotechnologies. They have a working definition of nanomaterial, which textually states:

Health Canada considers any manufactured substance or product and any component material, ingredient, device, or structure to be nanomaterial if:

1 It is at or within the nanoscale in at least one external dimension, or has internal or surface structure at the nanoscale, or;

2 It is smaller or larger than the nanoscale in all dimensions and exhibits one or more nanoscale properties/phenomena.

3 For the purposes of this definition:

4 The term "nanoscale" means 1 to 100 nanometres, inclusive;

5 The term "nanoscale properties/phenomena" means properties which are attributable to size and their effects; these properties are distinguishable from the chemical or physical properties of individual atoms, individual molecules and bulk material; and,

6 The term "manufactured" includes engineering processes and the control of matter. [10]

Canada is also focused on regulating nanotechnologies applied to the textile industry [11]. These regulations aim to allow global nanotextile commercialization using internationally accepted standards (International Organization for Standardization) to enable development and regulation [11]. The diversity of nanotechnologies from food and cosmetics to textiles and agriculture highlight the diversity of the field and the importance of proper regulation, which will allow safe and effective use as well as global commercialization.

Canada and the United States are working to harmonize the regulatory framework for nanotechnologies under the US–Canada Regulatory Cooperation Council (RCC) Nanotechnology Initiative [12, 13]. The objective of this harmonization is to share information, regulatory approaches, and risk assessments of nanomaterials and technologies to reduce risk and promote consistency.

To date, definitions remain broad, and regulations are limited as the development of nanotechnology continues. The overall goal in North America is to ensure the safety of consumers while allowing innovation within the industry. Nanotechnology-specific regulations are still lacking; however, as the field continues to develop, there is hope that new regulations will develop alongside to ease control of these technologies, especially considering the diversity of the field.

10.3   Regulatory Normative in Europe

Moving onto the European continent, if the goal is to properly comprehend the regulations in place, regarding the management of nanomedicines, it would be helpful to understand the inner workings of the European Union (EU). The EU is comprised of 28 countries in an economical and political union that makes decisions, though "*a hybrid system of supranationalism and intergovernmentalism*" [14]. The three legislative organizations of the EU are: (1) The European Commission, which proposes and puts legislation in place, (2) The European Parliament, and (3) The Council of Ministers. Twenty to thirty years ago, the regulation of medicines and other medical products was done through each countries corresponding agency with them all creating a vast network of cooperation (albeit very complicated) [14, 15]. The first step to harmonization was taken in 2001 when Directive 2001/83/EC was put in place to consolidate all previous legislation relating to medicines [14]. This is known as the "mutual recognition procedure". Applications could now be made to a specific member state and if authorized, be recognized mutually with other members provided the original state produced an assessment report [14, 16]. In 2004, the European Medicines Agency (EMA), previously the European Medicines Evaluation Agency, was set up to create a unified organization of regulations and protect public health by "*assuming responsibilities for scientific evaluation, supervision, and safety monitoring of medicines developed by pharmaceutical companies for use in the EU*" [14, 15, 17]. EMA is structured into several divisions including the Human Medicines Development and Evaluation (HMDE) and the Patient Health Protection (PHP) and scientific committees including the Committee for Medicinal Products for Human Use (CHMP) and the Committee for Advanced Therapies (CAT) [14]. Thousands of experts now make up EMA, guaranteeing multidisciplinary perspectives and the highest scientific standards [15]. More specifically, CHMP even created a specialized group of nanomedicines experts to provide scientific advice for the safe approval of products [17]. The EU has been backing nanotechnology since the early 2000s when it declared it a Key Enabling Technology that would address unmet needs and provide economic relief [14, 15, 18]. The EU is unique in that there is an explicit commitment and strategy to support nanomedical innovation [15]. There is, in general, a very positive and progressive perspective on the field in the EU and it is estimated that up to one-third of all products under development are now biotechnological in nature [14, 16].

Besides EMA, there are other agencies to which nanomedicines fall under their scope: the European Chemicals Agency (ECHA), the European Food Safety Authority, the European Environment Agency (EEA) and the European Agency for Safety Health at Work (EU-OSHA) [14]. While they each have their role, ECHA is worth discussing because of their establishment of the Registration, Evaluation, Authorization, and Restriction of Chemical Substances (REACH) legislation in 2006 (Regulation (EC) No 1907/2006) which centralized the regulation of chemicals in the EU [14]. REACH's definition of a chemical substance puts nano-compounds under its umbrella [16, 19] and the agency is charged with assuring high human health and environmental safety standards; they consequently evaluate chemical substances and their properties before they are put on the market [20]. For a detailed timeline of all the relevant events in EU legislation, please see [14].

To date, there are no specific nanomedical legislations in the EU. These technologies are currently dealt with using legislation on medical products, advanced therapies, and tissue engineering [16]. Additionally, the framework set by the EMA is aimed at making these regulations more specific to nanotechnologies. This is a regulatory approach in which the EMA and its subset agencies issue regulatory recommendations based on studies conducted on the technology in question [18]. These assessments are made publicly available, and the EMA has so far evaluated numerous marketing applications for nanomedicines [17]. Even if these "guidance documents" are not legally binding, they reflect EMA's interpretation of the legislation about specific nanotechnologies and the approval criteria required. Therefore, "the instructions and recommendations explained in the documents can be interpreted as compulsory in practice," and EMA's most active producer of nanomedicine guidance documentation to date is CHMP [14]. This framework has proven suitable, but there is growing consensus within the scientific community that something more robust is needed as the field continues to evolve [15, 21].

Similar to other regions of the world, one of the reasons for the lack of specific regulations towards bionanomaterials is the absence of a universally accepted definition [15, 18]. There have been many initiatives taken by the European Commission, EMA, and other agencies [14]. For instance, the European Commission put together the following: *"Nanomaterial means a natural, incidental or manufactured material containing particles, in an unbound state or as an aggregate or as an agglomerate and where, for 50% or more of the particles in the number size distribution, one or more external dimensions is in the size range 1 nm–100 nm"* [22]. Since this only helps determine whether a material is a nanomaterial or not (we cannot decide whether or not it is hazardous or safe) it has not been considered for incorporation into legislation as of yet [18]. Many experts such as Pita et al. believe that a working definition is needed to *"group materials traversing through different fields (e.g., medical, cosmetics, food, industry) into a defined class to properly account for their characterization, toxicity, and environmental risk assessment"* [15]. For a more descriptive explanation of the numerous definitions within the EU, please see [23]. Another source for the impediment of more robust regulations is the distinction between a medical product and a medical device which both have distinct methods of regulation. Products will usually act pharmacologically, immunologically, or metabolically, while devices will act physically [16]. Nonetheless, the distinction can become quite complex with nanomedicines that are increasingly blurring the lines and challenging current standards [15–17]. A reflection paper released by CHMP in 2006 recognized that bionanomaterials could cross-regulatory boundaries and insisted that "*appropriate expertise and guidelines will be needed for the evalua-tion of the quality, safety, efficacy and risk-management of nanomedicinal products*" [16]. D'silva et al. [16] offer compelling examples of these regulatory challenges via a case study approach.

By all means, the agencies of the EU have not stood idle in the face of these obstacles. EMA, in particular, has been cooperating with international counterparts to exchange information and promote global harmonization. In 2010, EMA hosted the First International Workshop on Nanomedicines with 200 European and International participants from 27 countries being present [15]. The ideas and allies formed during these back-and-forths later influenced EMA's direction in nanomedicine assessment [14]. Member states have also been contributing to the discussion and in 2012, France took the initiative and brought forward a national decree for the mandatory reporting of biomaterials "*to improve the knowledge of these substances and their uses, to ensure the traceability of sectors using these substances, to improve the knowledge of the market and the volumes sold, and to obtain available information on their toxicological and ecotoxicological characteristics*" [14]. Numerous countries such as Belgium and Norway have now followed suit [14].

10.4   Regulatory Normative in Japan

Japan has an advanced culture based on technology understanding, which also applies to how they have looked at nanotechnology. Thus, besides historical facts, two significant successes, related to nanotechnology, have emerged from this country. During the 1974 International Conference on *Production Engineering*, Norio Taniguchi was the first scientist to mint the term nanotechnology [24]; nonetheless, due to his contributions, Feynman is considered as the father of the field. On the other hand, another breakthrough came from a group of Japanese scientists led by Sumio Iijima in 1991 with the discovery of carbon nanotubes [25]. While these are a couple of historic landmarks, Japan has also had a considerable contribution to the field, with +3700 articles published only during 2017, and a variety of 407 products on the market involving electronics, healthcare, energy, food, among others [26]. Actually, under Japanese law, the definition of nanomaterial is "*materials in a solid state, which are manufactured using elements or other raw materials, and which are either nano-objects, or nanostructured materials (including objects that have nanoscale structures inside, as well as aggregations of nano-objects), with at least one of the three dimensions smaller than 100 nanometers*" [27].

Japan has two governmental agencies that have dedicated, part of their function, to the nanotechnology development: the Ministry of Education, Culture, Sports, Science and Technology (MEXT), and the Ministry of International Trade and Industry (MITI). While these have developed projects in the field since the 80s; it was not until 2000, when the Japanese government became aware of other countries initiative to enhance and regulate the nanotechnology field and the potential derivate businesses, that the country began to take bigger strides [28]. Therefore, and almost in parallel with the USA at the time, the Japanese Academic Association established the Framework Plan of Nanotechnology Research under the hood of the MITI in 2001 [29]. With their creation, nanotechnology was included as one of the four priority areas of development and investment in the next five years, as part of the Second Science and Technology Basic Plan, also known as STBP (2001–2005). Following the success of this initiative, the nanotechnology field has continued being part of more recent plans. The STBP is based on the Science and Technology Basic Law (1995), which "*aims to comprehensively and systematically advance science and technology policy. Thus, the government formulates the basic plan based on anticipating the next decade, putting into effect science and technology policies over a 5-year period*" [30]. Japan is actually at its 5th STBP (2016–2021), which not only has aggressive expectations to push numbers further but also goes deep in the development of big data, artificial intelligence, bio- and nanotechnology, being the final goal; this initiative of the government is to promote a "super smarter society" or "Society 5.0" [31]. In terms of the development sector, Japan has established nanotechnology clusters composed mainly of universities, research institutes, and companies within the field. Particularly in the case of universities and research institutions, these allow the cooperation with national and international companies, through the access to equipment and qualified personnel [32]. Currently, five active regulatory bodies are in charge in the management of nanotechnology's research and development: The Cabinet Office, New Energy and Industrial Technology Development Organization (NEDO), Japan External Trade Organization (JETRO), MEXT, and MITI.

While the government made a significant investment in R&D, the same was not happening with safety regulations related to the manipulation of nanomaterials. It was not until 2008, when, after studies that demonstrated unfavorable effects in rodents exposed to nanomaterials, that the Ministry of Health, Labor and Welfare (MHLW) created and disseminated the Notification on Present Preventive Measures against Exposure at Workplaces Manufacturing and Handling Nanomaterials after several meetings with expert panels [27]. In this report, eight main nanomaterials were identified as relevant for the country: carbon black, silica, titanium oxide, zinc oxide, single- and multi-walled carbon nanotube, fullerene, and dendrimers. The actual regulatory notification established several points that must be followed. While it is not the focus of this chapter to discuss these, the following points summarize the main characteristics of each notification:

Targeted Nanomaterial: Identifies any material that fulfills the definition of nanomaterials initially described in this section.

Targeted Operations: Refers to the manufacturing and handling of either nanomaterials or materials containing them, involving processes as disposal or recycling, and other non-routine operations.

Preventing Measures Against Exposure: This point is composed of several essential features.

Basic Concepts: Invites to the production sector and in-charge staff to educate their workforce before exposing them to the nanomaterial handling.

Investigation on Nanomaterials: The workplaces must ensure the availability of relevant information related to the nanomaterials, such as electron micrographs, size, etc.

Working Environment Management: This includes the enclosure of manufacturing and handling equipment, installation of a local exhaust ventilation system, dust removal measures, and periodic examination of nanomaterial concentration.

Work control: Consisting of preparations for operation rules such as cleaning floors, contamination prevention, and use of protective equipment.

Healthcare: It is mandatory for the employer to perform regular health examinations to the workers.

Safety and Health Education: Workers must be knowledgeable with the nanomaterial that they manipulate, the safety rules to follow, and protective gear to use.

Other measures: This involves the application of measures to prevent fire and explosions, and response to emergencies.

Dissemination of Information on Nanomaterials: To avoid risk of exposure to nanomaterials, selected workers, must be knowledgeable on the topic. Thus, it will be part of their function to educate other workers when handling specific nanomaterials.

Besides the regularly published reports that establish potential side effects prompted by nanomaterials, the Japanese government recognizes that further efforts must be performed to define any new legislative enforcement measures. Since, in the case of Japan who has not applied any modification to the MHLW notification since 2009. However, it is expected that the respective authority in the country will soon post a new regulatory normative.

The aforementioned is derived from an economy that historically has pushed the boundaries of sciences and technology, and as expected, the field of nanomaterials has not been left behind. To note that, in terms of demographics, Japan's people account for approximately 127 million inhabitants, according to a 2015 census [33], in a territory of $\approx$378 km^2 [34]. Further, Japan's gross domestic product (GDP) was close to 4872 trillion USD on 2017, with around 38,428 USD per capita income, placing the country as one of the top five most powerful economies worldwide [35]. Notably, in 2010, the nanotechnology businesses reached a market size of around 30 billion euros, with an expectations to increases this value sixfold by 2030 (numbers estimated by the METI) [32]. While it is expected that this increment in the market size could impact positively on the country's economy, the challenges and future directions that the Japanese society will face in the upcoming years, related to the development of nanomaterials, will be numerous. The Center for Research and Development Strategy (CRDS-Japan), has rigorously investigated and exposed these challenges, which includes topics of environment, energy, social infrastructure, life sciences, health care, information, communications, electronics, and design and control for materials and functions [28] (Table 10.1).

Table 10.1 Summary of the regulatory landscapes in the major countries discussed in this chapter

Country	Definition	Main governing agency	Presence of specific legislation relating to nanomedicines?	Presence of guidelines for nanomedicines?
Canada	Manufactured material at or within the nanoscale (1–100 nm) in minimum one dimension and has properties unique to nanomaterials that differ from the bulk material and/or atoms [10]	Health Canada	None	Yes
United States	Materials that have structural dimensions at the nanoscale (1–100 nm) (EPA) [9]	FDA	None	Yes
EU countries	Nanomaterial means a natural, incidental or manufactured material containing particles, in an unbound state or as an aggregate or as an agglomerate and where, for 50% or more of the particles in the number size distribution, one or more external dimensions is in the size range 1–100 nm [22]	EMA	None	Yes
Japan	Materials in a solid state, which are manufactured using elements or other raw materials, and which are either nano-objects, or nanostructured materials (including objects that have nanoscale structures inside, as well as aggregations of nano-objects), with at least one of the three dimensions smaller than 100 nm [27]	MEXT and MITI	None	Yes

10.5 Regulatory Normative in Emerging Countries

Following the USA's National Nanotechnology Initiative in the early 2000s, many of the emerging countries shadowed their example and are now at the tipping point between basic discovery, and commercial products and applications [36].

In Africa, South Africa and its government have *"made extensive investment toward creating a critical mass of infrastructure, equipment, and human capital for nanotechnology research"* through the publication of their National Nanotechnology Strategy (NNS) [37]. Their Department of Science and Technology (DST) also offer guidelines and is making a push towards nanotechnology-based rapid diagnostic kits for tuberculosis (TB) and human immunodeficiency virus (HIV) [37].

In South America, Brazil has held numerous workshops which led to the creation of the Renanosoma network, a consortium of organizations aimed at researching the potential effects of nanotechnology while increasing public awareness [38]. However, the push for innovative regulatory initiatives remains a challenge [36].

In Asia, South Korea has arguably produced *"the biggest commercial return"* thanks to their nanotechnology infrastructure [36]. As well, Isreal has been able to create many commercial products [36]. But the two leading countries coming out of the continent in this category are China and India. China is becoming a massive global player and is the world leader in nanotechnology literature production; they are also experts in nanocoatings and anti-corrosive nano-paints, among others [39]. The number of patents filed has overtaken those submitted by foreign bodies reflecting their push for commercialization [36, 39]. Similar to REACH, China put in place the Chemical Registration Centre of the Ministry of Environmental Protection (CRC-MEP) while nanotechnology standards are overseen by the National Nanotechnology Standardization Technical Committee and projects are overseen by the National Steering Committee for Nanoscience and Nanotechnology (NSCNN) [39]. It remains to be seen whether they will experience commercial success or not. In India, while there are still no laws governing nanomaterials, many governmental bodies such as their DST have created, and funded nanomedicine projects (Nanomission program) and the immense technological potential has been realized, albeit there are still concerns regarding safety and toxicity [40]. The country's scientists agree that there is a lack of a regulatory framework, and some have even gone so far as to offer proposals [41]. Currently, the Drugs and Cosmetic Act 1940 is what regulates drugs, devices, and diagnostics (including nanomaterials) with the DST continuously offering guidelines [40, 42]. Please see [41] for an analysis of the current specific statutes and guidelines in place that apply to nanomedicine. While India may not be as far along in a regulatory sense, they have been able to accumulate public investments for innovative projects and they anticipate more products will reach the market within the next decade [36, 40].

10.6 Concluding Remarks

With the fast-paced nature of nanomedicine, the stress on traditional frameworks is being made apparent and, as Trisolino et al. mentions, highlights how the law can lag behind new exciting technologies, making it difficult to *"find a balance between innovation and safety"* [20]. But as scientists continue to push the envelop and call for change, countries are progressively evolving their regulatory standards.

Acknowledgements Dr. Alarcon thanks the Canadian Institutes of Health Research (CIHR), the Natural Sciences and Engineering Research Council of Canada (NSERC), Ministry of Economic Development, Job Creation and Trade for an Early Researcher Award, and the New Frontiers in Research Fund—Exploration for a research Grant. Ms. Lazurko thanks the Queen Elizabeth II Graduate Scholarships in Science and Technology for financial support. Dr. Ahumada thanks the CON-ICYT—FONDECYT (Iniciación en la Investigación) grant #11180616, and to FDP-Universidad Mayor grant #I-2019077.

Disclosures All authors have read and approved the final version.

References

1. Larsson S, Jansson M, Boholm Å. Expert stakeholders' perception of nanotechnology: risk, benefit, knowledge, and regulation. J Nanopart Res. 2019;21(3):57.
2. Corley EA, Scheufele DA, Hu Q. Of risks and regulations: how leading U.S. nanoscientists form policy stances about nanotechnology. J Nanopart Res. 2009;11(7):1573–85.
3. Bowman DM, Hodge GA. Nanotechnology: mapping the wild regulatory frontier. Futures. 2006;38(9):1060–73.
4. FDA's approach to regulation of nanotechnology products. https://www.fda.gov/science-research/nanotechnology-programs-fda/fdas-approach-regulation-nanotechnology-products (2018). Accessed 23 May 2019.
5. Nanotechnology task force. https://www.fda.gov/science-research/nanotechnology-programs-fda/nanotechnology-task-force (2019). Accessed 23 May 2019.
6. Nanotechnology task force report 2007. https://www.fda.gov/science-research/nanotechnology-programs-fda/nanotechnology-task-force-report-2007 (2019). Accessed 24 May 2019.
7. 2013 Nanotechnology regulatory science research plan. https://www.fda.gov/science-research/nanotechnology-programs-fda/2013-nanotechnology-regulatory-science-research-plan (2013). Accessed 23 May 2019.
8. Research on nanomaterials. https://www.epa.gov/chemical-research/research-nanomaterials (2019). Accessed 24 May 2019.
9. Control of nanoscale materials under the toxic substances control act. https://www.epa.gov/reviewing-new-chemicals-under-toxic-substances-control-act-tsca/control-nanoscale-materials-under (2019). Accessed 24 May 2019.
10. Policy statement on health Canada's working definition for nanomaterial. https://www.canada.ca/en/health-canada/services/science-research/reports-publications/nanomaterial/policy-statement-health-canada-working-definition.html (2019). Accessed 23 May 2019.
11. Nanomaterials and their applications in textiles—standards. https://www.ic.gc.ca/eic/site/textiles-textiles.nsf/eng/h_tx03226.html (2019). Accessed 23 May 2019.

12. Amenta V, Aschberger K, Arena M, Bouwmeester H, Botelho Moniz F, Brandhoff P, Gottardo S, Marvin HJP, Mech A, Quiros Pesudo L, et al. Regulatory aspects of nanotechnology in the agri/feed/food sector in EU and non-EU countries. Regul Toxicol Pharm. 2015;73(1):463–76.
13. Joint action plan for the Canada-United States regulatory cooperation council. https://www.canada.ca/en/treasury-board-secretariat/corporate/transparency/acts-regulations/canada-us-regulatory-cooperation-council/joint-action-plan.html#s4.5.2 (2019). Accessed 23 May 2019.
14. Vencken SF, Greene CM. A review of the regulatory framework for nanomedicines in the European Union. In: Grumezescu AM, editor. Elsevier Inc.; 2018. p. 641–79.
15. Pita R, Ehmann F, Papaluca M. Nanomedicines in the EU—regulatory overview. AAPS J. 2016;18(6):1576–82.
16. D'Silva J, Van Calster G. Taking temperature—a review of European union regulation in nanomedicine. Eur J Health Law. 2009;16(3):249–69.
17. Ehmann F, Sakai-Kato K, Duncan R, Pérez de la Ossa DH, Pita R, Vidal J-M, Kohli A, Tothfalusi L, Sanh A, Tinto S, et al. Next-generation nanomedicines and nanosimilars: EU regulators' initiatives relating to the development and evaluation of nanomedicines. Nanomedicine 2013;8(5):849–56.
18. Soares S, Sousa J, Pais A, Vitorino C. Nanomedicine: principles, properties, and regulatory issues. Front Chem. 2018;6:1–15.
19. Satterstrom FK, Arcuri ASA, Davis TA, Gulledge W, Foss Hansen S, Shafy Haraza MA, Kapustka L, Karkan D, Linkov I, Melkonyan M, et al. Considerations for implementation of manufactured nanomaterial policy and governance. In: Linkov I, Steevens J, editors. Dordrecht: Springer; 2009. p. 329–51.
20. Trisolino A. Nanomedicine: building a bridge between science and law. NanoEthics. 2014;8(2):141–63.
21. Teunenbroek TV, Baker J, Dijkzeul A. Towards a more effective and efficient governance and regulation of nanomaterials. Part Fibre Toxicol. 2017;14(54):1–5.
22. Canu IG, Schulte PA, Riediker M, Fatkhutdinova L, Bergamaschi E. Methodological, political and legal issues in the assessment of the effects of nanotechnology on human health. J Epidemiol Commun Health. 2018;72(2):148–53.
23. Kreyling WG, Semmler-Behnke M, Chaudhry Q. A complementary definition of nanomaterial. Nano Today. 2010;5(3):165–8.
24. Taniguchi N. On the basic concept of nano-technology. In: Proceedings of international conference on production engineering. London; 1974.
25. Iijima S. Helical microtubules of graphitic carbon. Nature. 1991;354(6348):56–8.
26. StatNano. Japan indicators. https://statnano.com/country/japan (2019). Accessed 25 May 2019.
27. Director General of Labour Standards Bureau M. Notification on precautionary measures for prevention of exposure etc. to and other effects of nanomaterials. In: Ministry of Health Law, editor. 2009.
28. Strategy CfRaD. Nanotechnology and materials R&D in Japan (2018): an overview and analysis. In: Unit NM, editor. 2018. p. 86.
29. Tolochko N. History of nanotechnology. In: Valeri NCB, Sae-Chul K, editors. Nanoscience and nanotechnologies. Encyclopedia of Life Support Systems (EOLSS). Oxford: Eolss Publishers; 2009.
30. Ministry of Education C, Sports, Science and Technology. Science and technology basic plan. http://www.mext.go.jp/en/policy/science_technology/lawandplan/title01/detail01/1375311.htm (2019). Accessed 28 May 2019.
31. Office C. The 5th science and technology basic plan. In: Office C, editor. 2016.
32. Žagar A. Nanotech cluster and industry landscape in Japan. 2014.
33. Japan So. 2015 Population census. In e-Stat https://www.e-stat.go.jp/en/stat-search/files?page=1&toukei=00200521&tstat=000001080615&second=1 (2019). Accessed 28 May 2019.
34. Worldometers. Japan population. https://www.worldometers.info/world-population/japan-population/ (2019). Accessed 29 May 2019.
35. Bank TW. Japan. In data https://data.worldbank.org/country/japan (2019). Accessed 29 May 2019.

36. Matteucci F, Giannantonio R, Calabi F, Agostiano A, Gigli G, Rossi M. Deployment and exploitation of nanotechnology nanomaterials and nanomedicine. In: AIP conference proceedings 2018; 1990.
37. Dube A, Ebrahim N. The nanomedicine landscape of South Africa. Nanotechnol Rev. 2017;6(4):339–44.
38. Arcuri ASA, Grossi MGL, Pinto VRS, Rinaldi A, Pinto AC. Developing strategies in Brazil to manage the emerging nanotechnology and its associated risks. In: Linkov I, Steevens J, editors. Dordrecht: Springer; 2009. p. 299–309.
39. Jarvis SL, Richmond N. Regulation and governance of nanotechnology in China: regulatory challenges and effectiveness. Eur J Law Technol. 2011;2(3).
40. Bhatia P, Chugh A. A multilevel governance framework for regulation of nanomedicine in India. Nanotechnol Rev. 2017;6(4):373–82.
41. Bhatia P, Vasaikar S, Wali A. A landscape of nanomedicine innovations in India. Nanotechnol Rev. 2018;7(2):131–48.
42. Mishra M, Dashora K, Srivastava A, Fasake VD, Nag RH. Prospects, challenges and need for regulation of nanotechnology with special reference to India. Ecotoxicol Environ Saf. 2019;171:677–82.

Printed in the United States
By Bookmasters